Continental Drift
Colliding continents, converging cultures

Continental Drift

Colliding continents, converging cultures

Constantin Roman

MA (Bucharest), PhD (Cambridge),
Professor *Honoris Causa* (University of Bucharest)

Institute of Physics Publishing
Bristol and Philadelphia

British Library Cataloguing-in-Publication Data
A catalogue record for this book is available from the British Library.

ISBN 0 7503 0686 6

Library of Congress Cataloging-in-Publication Data

Roman, Constantin.
   Continental drift : colliding continents, converging cultures / Constantin Roman.
      p. cm.
   Includes bibliographical references.
   ISBN 0-7503-0686-6 (alk. paper)
   1. Roman, Constantin. 2. Geophysicists--Romania--Biography. 3. Plate tectonics. I. Title.

QE22.R66 R66 2000
551'.092--dc21
[B]

99-049926

Production Editor: Elizabeth Martin
Production Control: Sarah Plenty and Jenny Troyano
Commissioning Editor: Gillian Lindsey
Editorial Assistant: Victoria Le Billon
Cover Design: Jeremy Stephens
Marketing Executive: Colin Fenton

Published by Institute of Physics Publishing, wholly owned by The Institute of Physics, London

Institute of Physics Publishing, Dirac House, Temple Back, Bristol BS1 6BE, UK

US Office: Institute of Physics Publishing, Suite 1035, The Public Ledger Building, 150 South Independence Mall West, Philadelphia, PA 19106, USA

Typeset in LaTeX using the IOP Bookmaker Macros
Printed in the UK by Bookcraft, Midsomer Norton, Somerset

Dedications

To Teddy
My Mentor
And erstwhile Supervisor
[Sir Edward Crisp BULLARD (1907–1980)]
In Grateful Memory
This journal is dedicated by
His last PhD student in Cambridge

To my irreplaceable ancestors and siblings
The expropriated, the dispossessed,
The idealists of lost causes
And to my cherished successors
Without whose loving care
This outcome would have been impossible

Epigraph

It is as difficult to write a good life
as it is to live one.

Lytton Strachey

Contents

Author Biography

Professor Constantin Roman

Constantin Roman has an MA in geophysics, a PhD from Cambridge and is Professor *Honoris Causa* of the University of Bucharest. In addition, he is an international energy adviser and a Personal Adviser of the President of Romania. He started his career in offshore exploration at the time of the major oil discoveries in the North Sea. With the second oil crisis of 1978 he became an independent consultant working in the Norwegian and Barents Seas. Professor Roman's worldwide professional travels have brought him into contact with a number of extraordinary people and opportunities which allowed him to indulge his artistic pursuits. He has three children and lives near Glyndebourne where he cultivates his arboretum.

Foreword

Continental Drift is a book about universal values which transcend national frontiers or the confines of science and arts. Its author is part of that defiant species of uprooted who have chosen the sadly exacting role of exile rather than the tortured compromise of survival in a totalitarian regime. Yet, in spite of it, or perhaps because of it, Constantin Roman has never forgotten his beginnings. This sets him apart as an eminent ambassador of his native Romania and, at the same time, as a refined observer of his adoptive country, of which we find ample proof in the pages of his narrative. After the fall of Ceauşescu, these qualities were rewarded in Romania, where he was made a Professor *Honoris Causa* and Personal Adviser to the President of Romania. This overdue acknowledgment was apparent from the outset to a host of distinguished British 'worthies', who had known Constantin since his student days in Cambridge and who championed the *Roman cause célèbre* as a just and excellant one. Most prominent among them was Lord Goodman, Master of New College Oxford. Arnold Goodman was impressed with Constantin as a young man of *'impeccable character and absolute obduracy, reflecting an attitude of mind which has clearly developed from strong moral factors'*. More to the point, Lord Goodman was persuaded that Constantin was *'clearly determined to belong here and make a significant contribution to our national life'*.

On reading *Continental Drift* I can say, without fear of contradiction, not only that Constantin has discharged himself brilliantly of these expectations but that he had the merit of bringing to plate tectonics new models and concepts (in the Carpathians and Central Asia) which are still valid today. In particular, his definition of plate boundaries in continental lithosphere and the introduction of the concept of *'non-rigid'* plates or *'buffer'* plates, which are now called *'continuums'*, are still widely used. Few paid attention to his iconoclastic publication in a popular science journal (*New Scientist*) and in a short letter to the *Geophysical Journal* in the early 1970s, but the concept has withstood the test of time. If Roman's subsequent career as exploration adviser to Shell, BP, Exxon and a myriad of other major oil companies made him a world-wide expert in basin analysis and a successful oil finder, this was to our great loss in Academia. Predictably, Constantin's geodynamic studies, much praised by his clients in industry, could not be published for proprietary commercial reasons. This does not make him any less remarkable, as suggested by his Cambridge supervisor and mentor, Sir Edward Bullard, when he enjoined: *'Of course, it must be a wrench for you to leave Cambridge and take on wider responsibilities. If you want at some future time to come back to academic work, then no doubt you will be able to'*. Perhaps Constantin had never forgotten Bullard's prophesy. He comes back now to the fold (with a flourish) as Honorary Consul in Cambridge. I hope that he also renews his natural links with Academia, where he truly belongs.

Continental Drift offered me a relaxing excellent read, full of humour, humanity, wisdom and good science way beyond the History of Science. This book is an Ode to the Joy of Freedom, of a kind celebrated in Enesco's *Rhapsodies*, or the cosmic vision of Brancusi's *Column of Infinity*: this is Constantin Roman's *Ninth Symphony*. I trust the reader will share with me the pleasures that I have derived from reading *Continental Drift*.

John F Dewey, FRS, FGS
Professor of Geology and Fellow of University College
University of Oxford

Acknowledgments

I wish to express my gratitude to a myriad of people: to the great and the good for their encouragement, as well as to the less good, without whose contradictory forces I would have never persisted.

Gisèle, Corvin and Vicentiu helped with the word processing, the web site and the candid feedback. Mrs Charles Adeane, Mr Michael Conolly and Mrs June Miller read early versions of the manuscript. Mr Nat Page and Lady Catherine Page read at least a dozen different synopses and chose the 'winning' version for the publishers. Mr Adrian Phillips, Ms Lucinda Aris, Mr Julian Loose, Dr Simon Mitton and Mr Tristan Evans offered dispassionate and much needed independent advice, which steeled my resolve to continue. From Institute of Physics Publishing I am indebted to my commissioning editor, Ms Gillian Lindsey, for her steady patience and enthusiasm and to her colleagues, who kept the momentum alive.

Preface

On reading *Continental Drift*, it is perhaps ironic to consider the twists and turns of Constantin Roman's career, where he was led to excel in domains in which at the beginning he had not enjoyed, to put it mildly, an auspicious start. Constantin and I were contemporaries at the University of Bucharest, where we both read geophysics. At that time, I remember vividly, Constantin confided in me that when he was 16 he 'was not brilliant at either geology or physics', only to end up reading for an MA in geophysics. Furthermore, as a university student, it was immediately apparent that Constantin Roman's forte was most decidedly neither tectonics nor indeed seismology! Yet, at Cambridge, the topic in which he made his mark was 'seismo-tectonics'. This success is undoubtedly qualified by two traits of character acknowledged by his professors in Romania, namely perseverance and enthusiasm. To these I should add a third, which is crucial in our profession, that is imagination, which Constantin put to good use in interpreting his research data and coming up with unique solutions, often against all odds.

Apart from an introductory chapter about his Romanian roots (The DNA signature) and the period spent in Newcastle and Paris, in 1968–1969, this is a book of recollections of the author's time at Cambridge between 1969 and 1973 where he was a research scholar at Peterhouse. He was lucky to work on plate tectonics when this subject was in its infancy, as his supervisor and mentor, Sir Edward Bullard, led him to follow a path where each researcher was conspicuous and his scientific inroads significant. Now this same road is rather well trodden by a mass of individuals vying for prominence.

At Cambridge, this Romanian student was busy finding a solution to the occurrence of seismicity in the Carpathians and central Asia, which eventually led to a new definition of lithospheric plates. This new tectonic solution to the continental crust of Eurasia represented an early step in the development of plate tectonics theory and is unique in several ways.

First, and foremost, there is its scientific interest in the recognition of the existence of a new type of lithospheric plate—the *non-rigid plate* or *buffer plate*, published in various scientific journals. Several newly defined buffer plates were carved out of large tracts of the continental crust of Eurasia, in particular the areas behind the Himalayas—Tibet and Sinkiang. Further-more, the unexpected discovery of a then yet unknown piece of oceanic lithosphere, sinking vertically under the continental crust of the Carpathi-ans, represented a breakthrough in the reconstruction of the huge jigsaw puzzle of Tethys. The first results published in *Nature*, the *Geophysical Jour-nal* and *New Scientist* remain classics of the specialist literature. Although censored in Ceauşescu's Romania (yes, even science was emasculated for political reasons, as Roman belonged to the diaspora), Constantin Roman's work has endured the test of time and has the same topicality and freshness now as it had at the time of its conception. For this obvious reason, twenty-five years on, it was my privilege, as Scientific Director of the Romanian Geological Survey and an editor of the *Romanian Journal of Geophysics*, to publish *in extenso* Constantin Roman's PhD dissertation, 'Seismotectonics of the Carpathians and Central Asia' (*Romanian Journal of Geophysics* **18** 196, Bucharest 1998).

Constantin Roman's research was carried out at Cambridge under two remarkable scientists of world repute: first under Dan McKenzie and then under Sir Edward Bullard, himself remembered for the first-ever mathematical model of the Atlantic reconstruction, known as *Bullard's Fit*. As a pupil of Bullard, Roman's name falls within a direct line of distin-guished scientists of the Cambridge School of Physics, through Thomson, Rutherford and Cavendish, all the way back to Sir Isaac Newton. By pub-lishing today in Romania Constantin Roman's thesis, we acknowledge, albeit belatedly, the enduring character of this pioneering contribution to alpine plate tectonics. For these very reasons we welcome as a timely meet-ing of minds the parallel publication in England, by Institute of Physics Publishing, of *Continental Drift*, as an anecdotal history of the genesis of this same research. This makes for a perfect symmetry, as the two publi-cations in England and Romania complement each other.

The mid 1960s to early 1970s, which is the period of these mem-oirs, were the early but heady days when Vine and Matthews evolved the concept of 'sea-floor spreading' and the Canadian Tuzo Wilson, then a vis-iting Professor at Cambridge, devised the dynamics of 'transform faults'. It caused a frenzy of research which has since transformed geology in a manner which has not been done before or since.

On turning the pages of this story, the reader will gradually uncover the tensile forces beneath the real world of great scientists, with their frail-ties and their petty skirmishes, all leading to a climax which could not have been anticipated. This forms the backdrop to 'The Rat Race' chapter, a closely run contest, punctuated by youthful exuberance. The enthusi-asm paid off as, before the race was over, Constantin Roman lived through the beguiling excitement of beating a group of researchers from the Mas-sachusetts Institute of Technology to the answer to one of the great enigmas of Earth sciences—the seismicity of Central Asia. Cocooned in his Cam-bridge microcosm and obsessed by his research, Roman was completely oblivious of a trans-Atlantic team from MIT, working for years on the same problem as himself and gathering a wealth of information, which was about to be published. This sudden realization came as a shock, as the very object of the hard-won evidence which made the core of his Cambridge doctorate would have been put in jeopardy had the American team published their results first.

This is a unique instance when the reader can witness from within the researcher's camp his battles, trials and final triumphs. During the fray, many doubts were cast which inevitably confront every scientist. That is why, whilst in the throes of these struggles to the solution of a crucial scientific problem, one is never sure whether the pivotal new idea will be easily accepted by the geological profession, known more for its conservatism than for its innovative spirit or iconoclasm. Therefore, Constantin found it prudent to field these new ideas and test them against new audiences in a series of lectures, delivered as a guest speaker at British and Continental universities, in order to get them recognized before the dissertation was finished.

Somewhere in the remotest corners of academic etiquette there is an unwritten rule. This is the exclusivity of a field of research claimed by one's peers, which is generally observed by scholars within certain limits. It is in fact this very code of practice which had been broken whilst this story was unfolding. Once the ensuing 'rat race' was under way, there was no alternative but to go out and defend one's work, the paternity of which had to be preserved at all costs. This is how original concepts are born and what real science is all about—a drama often coloured by extraordinary ethos and raw feeling. The arguments, as they turned out to be at the outset, were often more twisted and the thinking of those behind them was more fractious than first anticipated. Indeed, as we know too well from the world of science, it is not uncommon for many a distinguished predecessor, or Nobel Prize winner at Cambridge and elsewhere, to live through identical crises.

Beyond the skirmishes of science, or scientists, *Continental Drift* is a book about the universal values of freedom, humanity, beauty and, above all, of *joie de vivre,* a song to the environment which inspires research and where enduring ideas are created. These are the impressions that nurtured this author's imagination and which form the very essence of the story, an interaction without which this work would not have been possible. They are the impact which Western Europe, and England in particular, had upon a fresh graduate from behind the Iron Curtain. On arriving in Newcastle, Constantin was youthfully unconcerned at having only five guineas in his pocket. Ironically, his travel ticket was paid for through a NATO grant, a source of monies which he had to keep secret from the Romanian authorities, lest his permission to travel to England might be withdrawn (NATO Secret). Under Ceaușescu, Constantin Roman was allowed out of Romania to deliver his scientific paper only after the conference had ended (!), a classic ploy in the armoury of Communist bureaucracy in its attempt to discourage contacts with the outside world, other than at official level. Worse still was to come, as a high-ranking Romanian official, and otherwise undercover agent for the Secret Services, tried to curtail Roman's attempt to do a PhD in the West. He labelled it a 'political option' (sic) and tried to discourage such an application with the bogus spectre of ending this exercise 'at best, as a mere waiter in a restaurant!'. Undaunted by such unsolicited prophecy, Constantin broadened his academic contacts and 'packed in the sights', as if there were no tomorrow. On the slow journey home to re-enter the Communist confines of his country, like any 'good Romanian', he stopped over in Paris, 'to see the Eiffel Tower'. In retrospect, I suspect, this move was only a subconscious excuse to meet Professor Thellier, a distinguished scientist of world repute and Head of the *Institut de Physique du Globe*. Thellier offered Roman a place for a doctorate in archaeomagnetism, but this was not to be, as the young Romanian arrived in France on 1 May 1968, only days before the Paris student riots plunged France into total chaos. With French academia in disarray, Roman's plans to study under Thellier came to nothing and soon he found himself stranded without money and without the possibility of returning to Romania, as his re-entry visa had expired. After some three months out on a limb, he was rescued from Paris by Professor Kenneth Creer's offer of a summer visiting studentship at the School of Physics in Newcastle. Here Constantin applied for and obtained a Research Scholarship from Peterhouse, the oldest Cambridge College, where he arrived in the autumn of 1969. Ironically, like Paris, Cambridge too was in turmoil, consumed by the 'Garden House riots' and Germaine Greer's *Female Eunuch*. This was only a feeble answer to and a copy-cat version of the Paris riots the preceding year, a kind of Guy Fawkes redivivus with a lot of fireworks. It was the time when Cambridge students set the 'Garden House Hotel' on fire and caused the Home Secretary to run for his life. Having been brought up in a dictatorship, where any form of dissent was instantly crushed, this Romanian student was utterly disconcerted to find himself parachuted, a nonplussed witness, into the middle of such unexpected events.

The contrast of cultures between East and West, between the author's preconceived, romantic and romanticized ideals of the West and real life, was always exhilarating, as he was led through encounters with eminent contemporaries in the worlds of Arts, Science and Politics ('Lotus-eater'). The people, the architecture and the gardens that surrounded him in his student days, and which formed a backdrop to his work, are remembered with a great deal of emotion and lyricism. If this scene is punctuated by irony and may be mixed with a good measure of Boswell-like frankness, I hope that the reader will forgive the author, as the intention was to present an unadulterated picture, as he saw it at that particular time. Opinions of those immature but blissful years are sometimes fraught with a youthful arrogance and therefore those of us who figure in these pages must read them in a compassionate spirit. Such emotive opinions could not be better expressed than in the words of Marie, Princess of Great Britain and Queen of Romania:

"Once I was a stranger to this people; now I am one of them, and, because I came from so far, better was I able to see them with their good qualities and with their defects" (*My Country*, 1916 Hodder & Stoughton).

In the above context it is nevertheless true that Constantin Roman's thinking, whilst it flourished in the stimulating Cambridge environment which represents the pinnacle of British academia, would not have been possible without the broad culture which he received from Romania. This confluence is reflected in the very spirit of *Continental Drift*. For, as we proceed, we must remember that this is not a textbook of popular science on the history of plate tectonics, but a series of personal impressions, or 'cameos', which some day might complement such a history of science.

As we turn the pages of this narrative, we see that Constantin Roman's singular road to Utopia was littered with disappointments and set-backs, as the darker side of human imperfections was gradually uncovered. However, these 'memoirs' are not intended as an exhaustive inventory of hardships, but rather as a quixotic refusal to accept them. This uncompromising stance is best summed up in the words of Thomas Mann: *"Finally, here, on Earth, there is only one problem left: how to get up! How to get up and go, break the chrysalis to become a butterfly"*.

On reading the book one may wish to mellow the edges, but one feels reluctant to do so, for fear of emasculating the complete and utter exhilaration with which such impressions were first recorded. They are an indispensable component of this narrative, which will benefit the reader.

Having considered the above caveats, one may well ask:

Could *Continental Drift* be a 'looking glass' wherein you could see yourself with the candid and unforgiving eye of the Continental 'drifting' within our midst?

Or, maybe, the resonance box of a musical instrument, which may amplify the 'drift' of this Continental author?

Or, could it rather be a history of science book, defining the drift of continents or the beginnings of plate tectonics theory?

At first glance all these three aspects may appear diverse, yet they have perfectly complementary and harmonious meanings, which should account for the triple entendre of the very title of *Continental Drift*.

Professor Sherban Veliciu
University of Bucharest
Scientific Director, Geological Survey of Romania
Vice-Chairman, Editorial Board, *Romanian Journal of Geophysics*

CHAPTER 1

THE DNA SIGNATURE

"Journeys, like artists, are born and not made. A thousand differing circumstances contribute to them, few of them willed or determined by the will—whatever we may think. They flower spontaneously out of the demands of our natures—and the best of them lead us not outwards in space, but inwards as well. Travel can be one of the most rewarding forms of introspection."

Lawrence Durrell (Bitter Lemons)

THE DNA SIGNATURE

Ever since Adam had a bite at the rotten apple, my ancestors have always made tactical errors, something to make them fall out with the establishment. Not that we came from a family who traced its lineage all the way to Adam, far from it: I came from a family where genealogical trees were scribbled on parchments, made of the skins of lamb foetuses. These scrolls were periodically burnt in the various wars that were waged, first eastwards and then westwards, wiping our lands both ways in the Carpathian foothills. Such cataclysms happened with monotonous regularity, like the ebb and flow of the ocean tide, until everything was wiped out of existence. In the end all that was left standing was the memory of bitterness, which eventually was purged by lost generations.

By the time I was born, woken from my mother's womb by the bombs dropped by Allied planes on their way to the Ploiesti oil fields, the family memory, which was not yet reduced to ashes, had been curtailed to the end of the 18th century on my father's side and to the Thirty Years War, at the beginning of the 17th century, on my mother's side.

This is not surprising, as women from the slopes of the Carpathian foothills had longer memories than men, as they had to remember with minute accuracy where they had buried the family treasures, hurriedly left behind in the wake of some barbarous invasion. These family trinkets and a few gold coins were assembled in haste in some earthenware pot

1

and hidden in a shallow trench. Often the secret was lost by the end of the war when the folk regained their burnt out hearth and it was not at all unusual for the oxen-driven plough, gently furrowing the rich soil of the family plot, to bring a pot of gold to the surface.

If the Wild West witnessed its gold rush, then each Romanian family had a story of a hidden treasure and to this day they are still looking for the pot of gold. Some ancestors went beyond the confines of their village in search of the lost treasure; foreigners in the new land of promise. More often than not my ancestors gave the treasure search up in despair and went instead to look for some long-lost ideal. They usually fought for the wrong cause, having espoused the wrong ideals. Then, as a result of their misjudgement, my forebears were driven out of their ancestral home. Often, they considered themselves very lucky indeed to be left alive, as they moved their womenfolk and children in wooden carts across the Carpathians to the relative safety of a neighbouring kingdom. Five generations ago, this was the case with Sava, the Transylvanian iron master of Fǎgǎras. In 1848 he busied himself sharpening pick axes for the revolutionaries set to rise against the oppression of the Austrian Emperor: he fought for the 'wrong cause' and had to flee some three hundred miles eastwards to the mouth of the Danube. There he married his daughter into another up-rooted family, who had fled the Catholic persecution after the 18th century partition of Poland, as these forebears were of the 'wrong religion'. These were my Roman ancestors, who were Moldavian merchants, of Orthodox faith, trading in Lemberg. As this Galician city changed hands with the change in political boundaries, it was ruled by the Catholic Emperor of Austria: the Roman family were under considerable pressure to renounce their Orthodox religion and convert to Catholicism, which they refused, paying the price of exile. They crossed the border into Moldavia, which was ruled by a prince of their faith.

The family history is full of such political and religious misfits, who preferred to take to the road, rather than compromise. Some of these idealists thought that they were asked to make a stark, if impossible, choice between right and wrong, or between black and white.

In reality the cause of displacement was simply the result of a clash of views of some very strong-headed people. One such 'strong head' was my great grandfather Venceslaus, a younger son of a younger son from Bohemia, who, on losing a court case over some family rights (against the whole village!) had his life made untenable. Little wonder that following this inauspicious legal event, Venceslaus resolved to migrate down the Danube, which he negotiated on a wooden raft, and eventually settled in Bucharest, in the middle of the 19th century.

It was here that this Czech great grandfather married another up-rooted person, a lady from Transylvania, who had the reputation of being a herbalist and a healer. She was called Ana and she came from a family of minor country squires from Eastern Transylvania, close to the source of the river Mures. Here, for generations, her folk built huge fortified churches on top of solitary hills, like pinnacles surrounded by curtain walls, within the precincts of which they would drive their cattle and store the grain to save it from the ravages of invasion.

Shortly before she died, Ana retired to her native Transylvanian village, which after the treaty of Versailles was no longer under Austria but integrated into the Kingdom of Romania.

Although I was born after this Transylvanian lady had died, I was fascinated by the family tales from this branch of the family to such an extent that, as a youngster of only fourteen, I went in search of my great grandmother's native roots. By that time, the village cemetery had been 'moved' and as the immediate family did not claim the grave, my great grandmother's remains were dispersed in a field of cultivated sunflowers. As a herbalist, I am certain that Ana would have approved of the change and as I looked incredulously at the field, I wondered from behind the face of which particular sunflower the old lady would be smiling at me?

There were some extraordinary tales passed down the generations from my Transylvanian ancestors and although I was born across the border, as it were, on the 'wrong side' of the Carpathians, I felt in a way that I was an honorary citizen of Transylvania.

That is why whenever I did not want to answer questions about my origins, which were all too often levelled at me, I would simply say that I came from Transylvania. This was not a technically true statement, but I always felt that I had some very strong claims to my roots in Transylvania.

Before the stories of the starving Romanian babies made the headlines of the British Press, Romania was not on the map of the British consciousness, yet many a Briton would nonetheless have heard of Transylvania.

"Does it really exist?" I would be asked in disbelief.

"Yes it does, I assure you."

"And what have you come here for?" my persistent torturer would always ask.

"Well, I have come here to put false fangs on the National Health Service", which would usually bring the conversation to an abrupt end, lest my bite prove more effective than my bark.

By the 20th century my family had settled down, or so it seemed. We were no longer in search of the pot of gold, nor were we in political or religious opposition. In the meantime, we came to realize that it was much safer to invest in an education, rather than in perishable property and heirlooms. So, we became a family of professionals, having abandoned the land which had for centuries nurtured so many generations, and reached the relative safety of the towns where we went to school.

After the Second World War, the latest political cataclysm shook my family, as Communism came to stay, for some forty years and Romania fell on the 'wrong side' of the Iron Curtain. My family then hoped to weather the storm by clinging to our education, the only asset which could not

3

(a)

(b)

(a) My mother Eugenia (Jenny) Velescu in 1928, when she was 16 years old and still a day girl at the Internatul Ortodox, the fashionable Orthodox Convent in Bucharest. Here Mother is wearing her first ball dress. Yes, girls knew how to have fun in Romanian convents—as Father was soon to find out for himself. (b) My father Valeriu Livovschi Roman in 1934 at the age of 26, sporting the famous dinner jacket which became unloved and unwanted under Communism, but was mothballed to be sent to Cambridge some 35 years on in 1969 (see section 'Gipsy's dinner jacket').

be taken away by the Communist régime, as everything else was either nationalized or confiscated, and all our savings were lost.

The yearning for travel was still in the blood and so was the persistence of clinging to the 'wrong ideals', an incurable habit passed on to me by many generations of uprooted and of dispossessed. That is why when, as a Romanian student in search of my Czech roots, I was seeking the family archives in Trebon, in Southern Bohemia, the archivist peered at me, behind his old spectacles and exclaimed:

"Young man, this is the call of the blood."

POSITIVE DISCRIMINATION

My first attempt at getting a passport was at the age of fourteen when, on receiving my first ID card, I immediately thought that it would automatically entitle me to obtain a passport: I had some Czech ancestry and was keen to discover my long lost relations in Czecho-Slovakia.

I went to the central police station in Bucharest and found myself in a room with many dejected elderly people, all of whom wanted to emigrate to Israel or America: being so young, I immediately attracted the attention

I was christened in the Orthodox faith of my forebears, in a private ceremony held at my family residence in Bucharest, in 1941. My Godfather was PPS to the Romanian Foreign Minister, Mr Manoilescu, who attended the private service and is seen here with my father Valeriu and my only sister Alexandra, then aged four. A few years later Mr Manoilescu was to die in the Communist prisons.

of the police officer, who asked what it was I wanted. I said I wanted a passport to travel to Prague.

"Are you travelling by yourself?"

To give greater weight to my request, I said that I would travel with my father, although he did not know anything about my initiative.

"All right then, ask your father to come here himself."

This was in 1955. I was in my early teens and felt that my world had fallen apart. I left the police HQ in sombre mood.

I scanned quickly, in the recesses of my mind, the virtues of our social pedigree, to see what chances I might have of being granted the freedom of travelling abroad, as passports were granted on stringent political and social class appurtenance criteria: clearly we were not born revolutionaries. Far from being Communist nomenklaturists our family did not want to compromise by jumping on the Communist bandwagon—quite the contrary they lost all their hard-won savings, their houses, business and chattels, they were marginalized. Our chances of survival were not very good, let alone the luxury of being granted a passport.

My mother Eugenia (Jenny) Velescu was born in Bucharest, in 1912, and came from a prominent professional family. She was the youngest daughter of George Velescu and Ana Zelişka. Her father, George, graduated in law and pharmacy and chose to profess the latter, on the advice of his close friend, His Beatitude the Patriarch Miron Cristea, Head of the Romanian Orthodox Church. Grandfather George started his career at Bruss, by appointment pharmacist to HM King Carol of Romania, and was decorated, in 1906, for his services to the King. Subsequently my grandfather started a successful pharmacy practice and eventually rose to prominence to become President of the Romanian Pharmacists' Association, and Editor of the *Pharmacopea Romana* and the *Curierul Farmaceutic* magazine in Bucharest, and became sole distributor of Merck's drugs in Romania.

My mother's mother, Ana, was of Bohemian (Czech) and Transylvanian (Szekler) parentage. She was born in Bucharest where she was educated at the Bărăţia Catholic Convent. Grandmother Ana spoke four languages fluently, was one of the first women in Romania to graduate (in violin and piano) from the Music Conservatoire in Bucharest and during the First World War took a degree in pharmacy in order to be allowed to run her husband's practice during his absence on military duty. As a child I vividly remember my grandmother's huge art collection and library and until the advent of Communism, when family properties were expropriated, I attended Granny Ana's chamber music recitals held on Sunday afternoons at her home in Bucharest. Ana's personality was to have the greatest influence on my upbringing, first in learning foreign languages and also in opening my interest in arts and science.

Unlike her parents, Mother was not an academic, but she was educated to converse in several foreign languages and play the piano. At her parents' home, in Bucharest, Mother was attended by servants, but when Communism came, she adapted manfully to the harsh conditions imposed on her and did not shrink from her responsibilities as wife and mother, in a climate of political oppression, fear of imprisonment and financially reduced circumstances.

My father, Valeriu Livovschi Roman, was born in Bucharest in 1906, the eldest son of Vicentiu Livovschi Roman, a pharmacist, and of Stefania Burada, only daughter of Rev. Constantin Burada. My grandfather Vicentiu came from a family who served the Romanian Orthodox Church for seven generations and had a strong musical tradition, which produced a composer of Russian liturgical music at the Tzar's cathedral in St Petersburg. On my paternal grandmother's side, the Buradas were a landed family of Orthodox clergy who were benefactors and founders of churches and schools in the 19th century, but also produced composers, judges and an anthropologist.

My father, Valeriu, graduated in industrial chemistry at the University of Bucharest, in 1930, and started his career in the chemistry laboratory of '*Phoenix*', an Anglo-Romanian oil company, with producing fields and a refinery at Ploiesti. Father was soon promoted to become a senior executive responsible for oil export, first at the Company's terminal at Constanta, a port on the Black Sea, and from 1937 at Giurgiu, a port on the Danube where I was born in 1941. When the Giurgiu terminal was bombed in 1941, my father moved to the company's HQ in Bucharest, where he became Head of Marketing, but not for long as the Communists nationalized *'Phoenix'* in 1948 and in the ensuing witch hunt Father was accused of being a 'collaborator' with the British. He escaped the Communist prisons only due to the strong backing of the company's workers, with whom he was very popular and who pleaded for his reprieve. In the process Father was made redundant. After several years as a chemical engineer in industry, his career ended as an editor at the Editura Technică publishing house in Bucharest. As he refused to join the ranks of the Communist Party, Father's name was not allowed to appear in print on the back cover of the many science books he produced, and was relegated instead to a relative professional obscurity. Still, in spite of these inauspicious circumstances, he managed to produce several patents in the field of catalysis of sodium chloride and wrote a classic book on the subject.

The political and social scene in Eastern Europe was not going to improve. Although after Stalin's death, in 1953, a slight relaxation occurred in Romania, with the Moscow-trained Communists being purged from power, by 1956 the Hungarian uprising against the Russian occupation troops and their Communist stooges caused a tremendous backlash throughout neighbouring Romania. I thought for a moment that we too would have been liberated of the Communist-led oppression, Russian in particular, so I slackened my studies of compulsory Russian language and nearly had to repeat the fifth form. I only did well at the subjects where I liked the teachers and neither physics nor geology were favourites of mine. We were taught geology by a female Communist Party activist, who knew nothing of the subject and who 'politicized' her classes, by telling us, literally, that:

"The Capitalist's over-production of hydrocarbons in Romania was criminal, as it did not allow for the oil and gas fields to regenerate."

"When, in our life time?" I would tease the teacher, but she could not understand the jibe. At the age of 15, when we started to be critical of all values taught and in particular of Communist values, we could not stop laughing and did not take geology seriously for that matter. For my ten A-level exams I merely scraped by, with geology being at the bottom limit. Besides, I had known ever since I was in primary education that I wanted to become an architect, for which geology was superfluous. The Russian repression of the Hungarian Uprising, in 1956, had a long-term negative effect on the selection process for university entry. This was a ruthless, positive discrimination, based on social class criteria. The immediate result was that by the time I had to sit the entrance tests for the School of Architecture, in 1959, only 20% of the places were allocated to sons of 'professionals', of whom I was one, as my father had a university degree and worked in a publishing house. Out of the available 60 places for the first year of the School of Architecture, therefore, I could only compete for a mere handful of 12 places, for which there were hundreds of candidates. The two drawing tests were crucial in being short-listed for the second tests of maths and physics. Quite apart from the intrinsic merits of the drawing tests each candidate was given a 'social weighting' mark as a function of his parents' Communist Party membership, nationalized property and the like. As no member of my family had joined the Communist Party and my family's houses and business were expropriated, I had a negative handicap from the outset. To counteract this inevitable disadvantage, I had prepared for two solid years, with private coaching every week in physics, maths and drawing. But all this was to no avail, as I could never ever be short-listed following the arts eliminatory tests in drawing for the School of Architecture. At this point, Father beseeched me to face up to the political reality, painful as it was, and agree that I had better go for admission exams in science, rather than in arts. It stood to reason that in science, at least, the results of the exams tests were unequivocal and not open to interpretation for political ends.

It was somehow ironic that with poor A-levels in geology and physics, I would sit the following year the admission exams for geophysics, at the Faculty of Geology in Bucharest. On this occasion there were only 5.5 candidates for one place, but as the equations provided unique answers I could prove my strength in the maths and physics tests. My father was relieved to see me through, as education had become a symbol of survival in a family where all savings were confiscated and all properties and chattels gone. So, in our family, as in many other professional families in Romania and throughout Eastern Europe, education became a symbol of resistance to the Communist system. This is how I became a geophysicist.

DRIFT TO GEOPHYSICS

To appease my disappointment at training to be a geophysicist, rather than an architect, Father tried to play up the *'cosmopolitan'* character of the school I joined, where many students came from Africa, the Middle East and South America. These were the aspiring Communist régimes, which would send students to study petroleum geology and drilling in Romania; all in all, an uninspiring motley group of assorted Syrians, Iraqis, Cubans, Algerians, Nigerians and Albanians, most of whom were Communist sympathizers, which I was definitely not! Besides, what separated the Romanian native student population from their foreign colleagues was the latter's freedom to travel abroad. By contrast, Romanian citizens had no automatic right to a passport, unless they were from the higher echelons of the Communist party. We resented our foreign colleagues returning from holidays with presents for their Romanian girlfriends, with whom they were very popular. Understandably, for this same reason, the foreign students carried less favour with the native male student population, who were at a definite disadvantage, for lack of cash and travel opportunities. Eventually the world politics of super-powers came to our rescue: the Cuban students were expelled from Romania for demonstrating, without permission, in front of the American Embassy in Bucharest, following the debacle at the Bay of Pigs. The Cubans were soon followed by another wave of expulsions of the Albanian students, this time, because their Government sought a rapprochement with China, at the expense of the Russians, who were still Romania's staunch allies. Soon after this clean up job, some of the North African Arab contingent withered away in disillusionment at seeing for themselves that Communism did *not* actually work in practice. So, the Romanian male students found themselves, overnight, masters of their own backyard over the bevy of rudderless girls whose boyfriends were unceremoniously kicked out of the country. We could hardly disguise our jubilant mood: in fact we flouted our hitherto repressed male chauvinism and our newly found self-esteem as unchallenged kings of the Communist castle. The rejoicing was short-lived, as most students had to keep their nose to the grindstone and make sure that they passed all exams by the end of each year.

The whole teaching structure was rather Victorian in practice, with an obsession for technical details, which we had to absorb regardless of their relevance. We had ten compulsory exams every year and in all sixty different courses. These ranged from maths, physics, chemistry, even combustion engineering, drilling, hydrocarbon geology, structural geology, tectonics, stratigraphy, mineralogy and crystallography, palaeontology, geochemistry, followed by gravity, magnetics, seismology, electric logging and a myriad of other courses, with a sprinkle of languages, economics and politics (Marxist ideology, of course). There were no published textbooks and we were relegated to taking notes during lectures, of which there were six a day plus practicals (laboratory work) and summer field trips.

Faced with what I considered to be a barrage of wanton, old-fashioned teaching I decided to be selective, an attitude which was frowned upon. I needed my space to develop personal interests, some of which were extra-curricular activities, like earning some much-needed money to buy myself basic clothes, which my father's meagre professional salary could not provide. Although the university education was free, I had no right to a student grant, under a rule of positive discrimination against the sons of professionals. However, some of the mature students, selected from amongst factory workers, did have a sponsorship from their original place of work. The academic going was tough and should we fail exams at the end of the first year, we would be asked to leave. As it happened, by the fourth academic year numbers dwindled from the original sixty to a mere forty or so students, who were still in the race. During the academic year I would earn a little extra cash doing abstracts of science papers from French and English journals, occasionally publishing articles in *Viaţa Studenţească*, the Bucharest student rag (travel fiction, interviews), or in high-brow national literary weeklies such as *Luceafărul* and *Contemporanul*. Writing travel fiction became an obsession and escapism, securing a minimum of sanity: as I would not be allowed to travel freely abroad I would do it with the eyes of my imagination and pretend that I did go and visit Copenhagen, Edinburgh, Palermo or London. So I wrote about the spires of Copenhagen, which I never saw, I invented a fictitious interview with Count Tomaso di Lampedusa, the Sicilian grandee, as an excuse for introducing to the Romanian public his best-seller *The Leopard*. I wrote about the Hogmanay revels in Edinburgh, about the University of Sussex, in order to introduce its architect, Sir Basil Spence, and 'travelled' to many more places I had never set foot into, but dreamt of visiting one day. In addition to having fun with such publications, I also had the secret pleasure of 'educating' the Romanian public, whilst I would get a bit of pocket money.

As for the summer months, when I had no practicals I took up the job of courier for the National Tourist Agency in Bucharest. This allowed me to visit the 16th century Moldavian painted churches, the castles of Transylvania or the beaches of the Black Sea resorts.

Contacts with foreign visitors from the West were expressly forbidden in Romania, and if they took place at all they had to be 'reported' to the police. In my summer job as a tourist guide I found a major loophole in allowing unfettered contacts with the 'free world' as well as the ability to practise my English, learn about the West, get some pocket money and have a free holiday which I could otherwise not afford.

Still, I was well aware of the fact that sundry hotel waiters, receptionists and coach drivers were informers of the dreaded Securitate, but I was too young and too reckless to care for such 'details', which eventually brought me into conflict with the authorities.

Through my courier activities I made friends abroad, who would subsequently send me much-needed dictionaries or foreign literature unavailable in Romanian book shops for economic or political reasons. In this way I managed to smuggle into Romania (and not without trouble) the 24 Penguin paperback volumes of Churchill's *History of the Second World War*. Should I have been caught in possession of such illicit political material I would have risked my freedom for this crazy enterprise and would have been expelled from university, for political and social dereliction, branded an 'enemy of the people'. Quite so and proud of it too!

The Romanian School of Geology had a long tradition of links with the German, Austrian, Belgian and French schools going back to the 18th and 19th centuries. This was due to the research on the mineral ore deposits of Transylvania, the salt mines and the oil and gas fields of the

In this photograph I am seen with this charming orthodox nun in Bukovina, northeastern Romania, at the convent of Moldoviţa, one of the five most famous 16th century churches preserving their external frescoes. My blue jeans and Marks and Spencer red jumper, received as a present from one of my former tourists, were in defiance of the Communist strictures which regarded western style outfits as 'decadent'. This was in effect a political statement, as I was proud to be 'decadent' (in the Communist sense of the word). I supposed even my charming companion in her black garb must have approved, as she allowed herself to make an allusion, in French, to Stendhal's novel Le Rouge et le Noir. I visited Moldoviţa 18 times, on each occasion with coach loads of forty or so tourists, and perhaps the superb frescoes as much as this attractive nun had a lot to do with it.

Carpathians. Gold and silver mines had existed since Roman times. Salt had been an export commodity since the Middle Ages, and oil seepages used since ancient times for lighting homes and oiling cartwheels. By the mid 19th century the oil industry started in earnest, at the same time as in the United States, with the oil fields of Ploiesti looking like a site from wild Texas during the oil rush. It was on the gas fields of Transylvania that Count Eötvös tested his gravity method of exploration. The terms 'diapir' and 'diapirism' were coined at the beginning of the 20th century by Ludovic Mrazek, a professor from Bucharest University, and it is in Romania that salt diapirs were demonstrated to provide an effective seal for hydrocarbons. Until then, geologists were only looking for classical dome structures, or 'four-way closures'. The new salt sealing concept broadened substantially the areas of investigation worldwide. It was also in Romania, in 1923, that two brothers, Conrad and Marcel Schlumberger, graduates of the Paris School of Mines, tested their electrical resistivity methods for hydrocarbon prospecting. The discovery of the Aricesti oil field, a salt-controlled structure, near Ploiesti, was a turning point in geophysical prospecting. The survey was commissioned by Jules Ménil, the President of the *'Steaua Română'* Oil Company. It took several years before the Schlumbergers managed to persuade the Texan oil companies about the advantages of electric prospecting.

The Schlumberger brothers were beckoned to Romania by their contemporary from the Paris School of Mines, Sabba Stefanescu, who put the mathematical foundations to some of the Schlumberger methods. Essentially, the whole idea was going to make famous the Schlumberger brothers and, subsequently, their business empire. They knew that different rock formations drilled in the process of oil search could be identified (lithology, thickness of formation, physical properties, etc) by sending a miniaturized apparatus down the drilling hole. This equipment would take physical measurements of the rock resistivity/conductivity through an induced current, which would register on a paper log the variations in the response of the rocks. These logs would be interpreted and the geology identified. The whole idea was to revolutionize oil exploration and make it more reliable and cost effective, because, prior to that, the only scant knowledge obtained about the subsurface geology was through the bits and pieces of rocks which came to the surface from the drilled material. This was not sufficient to know the depth from which such material came, the thickness of the beds, etc. The first visits to the United States, intended to demonstrate the logging method, were met with scepticism, if not indifference. In the America of the 1920s the prevailing philosophy in oil exploration was one of 'wildcatting', that is of drilling new exploration wells on the basis of risk-taking, rather than geological interpretation. It was the time of an upsurge in oil discoveries and understandably science was not needed to become a successful oilman. Besides, how was one going to demonstrate that the results were right or wrong?

The answer was to test the method on a known producing field, which was done in Romania. From then on the Schlumbergers never looked back. The method was established, patented and subsequently used throughout the world on every single oil well drilled. Of course, new methods were introduced to measure other physical parameters (magnetic properties, radioactivity, seismic response, etc) intended to define the porosity, permeability, oil or gas saturation, water saturation of the sub-surface formations. Not one log, but several such logs were registered and compared, which made the method more comprehensive and reliable.

Log analysis became a science in its own right, the bible of the oil geologist and an important course at the University of Bucharest, where I was
studying. Ştefănescu, later to become a Fellow of the Romanian Academy,
taught electrical prospecting at the School of Geology in Bucharest and he
was on the board of examiners of my Dissertation.

Amongst the plethora of sixty or so professors and lecturers we had
in Bucharest, the Head of Geophysics, Professor Liviu Constantinescu, was
odd for a variety of reasons. He combined his patrician-like demeanour
with Communist Party membership, where he was an active member. This
position allowed him access to a passport and foreign travel, generally denied to non-party members. Liviu Constantinescu spoke fluent French,
English, Russian and German and was appointed to many national and
international committees. This secured for him in addition the extraordinary advantage of regular foreign travel, contacts with western academics
and their research topics, as well as a number of subscriptions to foreign
scientific journals, which he kept locked up in his office. It is from Constantinescu that I learned for the first time about the methods of physics
applied to the history of art and archaeology at Aitken and Hall's Oxford
Laboratory, likewise about the archaeometrical research of the Lerici Foundation in Italy, or the magnetic research on archaeological pottery artefacts
of Professor Thellier in Paris. This was exciting stuff for me because it
brought art into science, or rather opened the perspective of bringing arts
back into my life through the 'back door'. Essentially, what Lerici did in
Italy was to introduce geophysical prospecting methods used in hydrocarbon exploration to the much finer and smaller scale of archaeology. The
same methodology used in finding buried oil structures was now applied
in discovering buried historical cities, Etruscan tombs, fortifications and so
on. Even more riveting was the under-water archaeology which identified,
off the coast of Egypt and Greece, sunken ships full of treasure trove. A
new world suddenly unfolded before me and I was entranced: geophysics
was not so dull after all!

It was Constantinescu who brought to our attention, during his lectures, the first elements of palaeomagnetic research carried out previously
at Cambridge and subsequently at the School of Physics in Newcastle by
Runcorn and Creer. Earlier on, in Paris in the 1940s, Professor Thellier established from measuring the magnetic intensity properties of fired bricks
from Carthage that their values were a function of age. As the bricks were
dated with great accuracy by the archaeologists, it emerged that the Earth's
magnetic field intensity varied in time. In practice, it was discovered that
as the clay was fired for the purpose of turning it into bricks, at high tem-
peratures in the kiln, all magnetic iron particles aligned themselves in the
prevailing direction of the Earth's magnetic field, like little compasses,
used in navigation. Once the temperature of the brick dropped and the
firing process finished the new orientation of the magnetic particles would

remain 'fossilized' and the whole brick would display the same azimuth as the Earth's magnetic field of the time. A standard curve was established from statistical measurements which allowed conversely to date bricks from sites which could not be dated through traditional archaeological methods, by simply plotting their total magnetic intensity on the standard curve.

The methodology was extended from historical to geological times and gradually the Earth's magnetic field variations were established for each continent for particular geological times. Once the palaeomagnetic variations of the Earth's field were known in space and time, it emerged that at different geological times continents had had different positions on the surface of the globe. The theory of 'continental drift', suggested by Alfred Wegener, at the turn of the century, had practically been ignored after the 1920s but now began to look much more promising. Wegener, a German astronomer turned meteorologist, noticed, as many had before him, that the curves of Africa and South America fitted together rather neatly. He compiled solid observational evidence, both biological and geological, for the concept that the continents had moved apart. His ideas were not accepted when he introduced them first in 1912 and later in his book *The Origins of Continents and Oceans*, not only because he was regarded as an outsider by the geological establishment, but also because he could not convincingly explain the mechanism of continental drift. This objection persisted until the 1970s and its main proponent was Sir Harold Jeffreys.

What Wegener's intuition suggested at the turn of the century, now became a scientific fact that could be demonstrated in much finer detail. 'Continental Drift' got a fresh impetus (the term plate tectonics was not coined until after 1965) and it was the buzzword amongst Western geoscientists. Behind the Iron Curtain, with the paucity of scientific information imposed by ideological censorship, continental drift had the same mysterious attraction as the Sphinx's riddle to which only Liviu Constantinescu had the answer. Such mystery spurred my curiosity and I soon decided to avail myself of my communication skills in foreign languages to write 'fan letters' to the main palaeomagnetism players in Newcastle and Paris. I also wrote to the Oxford laboratory, which dealt with methods of restoration, conservation and dating of archaeological artefacts. It was a pleasant surprise to receive reprints of scientific papers, which until recently were the sole preserve of Constantinescu. This was the beginning of a long and fruitful correspondence with my colleagues in Western Europe, which was eventually going to be my life raft, when stranded in England and France without a return visa to Romania.

FIRST PASSPORT
I waited another eight years to the age of 22, before I eventually attempted again to obtain a passport to travel to Poland, through Hungary and

Czechoslovakia. By this time I was a student in geophysics at the Institute for Oil, Gas and Geology and had many contacts with foreign tourists through my summer job as a courier. Even then the travel was not a straightforward affair. I had to produce to the passport authorities letters of invitation from Poland from my would-be hosts stating that they took entire responsibility for my upkeep during my visit to their country and that I would not ask for any foreign currency for my trip.

I went twice to Poland, Hungary and Czecho-Slovakia. These trips gave me a foretaste of the West, with the Gothic and Baroque churches of Prague, with the broad vistas of the Danube in Budapest, with the avant-garde music and paintings of Poland and French and English books and newspapers available there, but forbidden in Romania. The *Theatre of the Absurd* of my fellow countryman Eugène Ionesco, an exile in Paris, was not played in Romania, but I could enjoy it, for the first time, in Poland. Now I could read the books of Vintilá Horia, another uprooted Romanian and winner of the 1960 Goncourt prize for his deeply moving historical novel, *Dieu est né en exil*. A new world was opening up to me, a world I knew existed, but could only dream of, and I felt almost inebriated.

As for travelling to Western Europe, I knew that I would never be granted a passport. This was immensely frustrating, as my fluent knowledge of French and English gave me access to the latest Goncourt and Renaudot literary prizes in Paris, or the reviews of the Cannes and Venice Film Festivals. I was longing to see for myself the Impressionist paintings in the Courtauld Institute in London and the Orangery in Paris, the churches of Christopher Wren, the Oxford colleges, the modern architecture of Le Corbusier and Sir Basil Spence, and listen to the rock music of the Beatles.

PALAEOMAGNETIC LINK

My fledgling contacts with researchers in Oxford, Newcastle and Paris who worked on palaeomagnetism grew stronger. I found the subject fascinating especially for its applications to dating of archaeological clay artefacts. Oxford's Laboratory of Physics applied to the history of art and archaeology seemed even more riveting. We had no such facilities at our disposal in Romania. I absorbed ravenously the contents of the scientific paper reprints which were sent to me from abroad and within a short period of time I had enough confidence to request that my diploma dissertation should be in palaeomagnetism. This sounds an easier choice than it seemed at the time, as it went against the grain of the established system: in effect the Head of the Department had devised forty or so different topics of research, one for each student in his final year. We were supposed to choose the subject in the order of our exam success. I was towards the bottom of this league, for my unorthodox and much frowned upon practice of being selective in my performance, but I still had to jump through the hoops. I found little choice in Constantinescu's list to fire my imagination for a proposed MA dissertation. I said that I wanted none of the topics on the list and I preferred instead a subject in palaeomagnetism. Far from being pleased about the enthusiasm inspired by one of his own lectures, the professor retorted that I could do as I pleased, but that he could not condone it and would not guarantee success. In fact, he used this excuse to wash his hands of all responsibility in what appeared to be an unconventional choice.

Undaunted by this inauspicious beginning, but pleased enough about the green light, with all its gloomy caveats, I contacted the only specialist on the subject at a geophysics laboratory in Bucharest and asked for help. I was soon off to a fresh start on the suggested topic of the 'Palaeomagnetic properties of the mineral ores in a copper deposit of the Dobrogea district'. I went on site to collect samples for my measurements and here a skilled worker, a former political prisoner who still suffered a 'forced domicile', accompanied me down the mine. This was in 1965, a period of relative political relaxation at the beginning of the Ceauşescu régime. Although I considered myself politically and socially aware about the repression suffered by the Romanian professional classes at the hands of the Communist dictatorship, I was numb at the stories I was told by my newly-found companion. I also lived the frustration of being totally unable to do anything to improve his plight, other than to listen sympathetically and not divulge his confession. A strong complicity developed between us, which made my odd stay in this mining community an enriching experience. I took the rock samples to Bucharest and started to prepare the specimens for the measurement of remanent magnetism. The brief I had was to comment on the results from two groups of samples from the actual mineral ore and to compare them with similar measurements from samples taken from the sterile rock formation. The study was intended to settle a long-drawn-out dispute between two schools of geological thought, regarding the epigenetic or syngenetic origin of the copper ore deposits. This had a practical implication in the future development of the mine. It was easy to understand that, whichever way the results would tilt the balance, the conclusions would be quite exciting and I plunged myself with great enthusiasm into crunching the numbers and integrating the palaeomagnetic and geomorphologic field data.

POTENTIAL DEFECTOR
The first results of my palaeomagnetic research seemed encouraging enough to present them to a Congress of Geology in Belgrade: I was denied access to a passport and consequently was not allowed to present the paper myself. Yugoslavia had an open border with Trieste and the Romanian apparatchiks, mindful of the possibility of defection to the West, decided that I was not sufficiently 'reliable' to be allowed to travel. The refusal came as no great surprise, but I was not to be deterred. If anything, this

'no confidence' vote made me even more determined to try again.

The *Proceedings of the 9th Carpathian–Balkan Geological Congress*, in Belgrade, published my article in 1967: I sent a reprint to Professor Runcorn at the University of Newcastle, one to Professor Thellier in Paris and another to Oxford to Teddy Hall. Soon I had a new article published by the *Geophysical Journal of the Romanian Academy* and I was glad I could reciprocate the exchange of reprints with my western colleagues, even though I was aware that my contribution was quite modest.

The idea of visiting the specialist university laboratories in France and England started to germinate in my mind and so I arranged to receive an invitation to England from a young British friend. I had no great hope in succeeding in an enterprise in which I had already failed in the previous year, but I felt it was certainly worth trying again. In the event I was inevitably refused. At my suggestion, my English friend tried intervening through his MP whom he asked to impress on the Romanians to grant me a passport—it was a long shot and it did not work. This was 1966. I was 25 and already my request for a passport had been refused three times.

The same summer of 1966 I had my finals and part of these involved the writing of the MA dissertation and an oral presentation to a panel of examiners. To this end I already had several scientific articles to my name, including the Belgrade paper and several articles on physics applied to archaeology. I also took recorded interviews from known geologists presenting contradictory views on the genetics of the copper mine deposits. This put me in the invidious position of being a kind of scientific referee with an undisputed new method of deciding the 'truth' and this fired me with a youthful enthusiasm. I should have got 10 out of 10 for my work, but, mindful of my academic past, the examiners could not come to terms with this success and marked me with 9/10. This represented 50% of the whole score with the other 50% being the average of the 60 odd exam results taken over the previous five years. Furthermore, the examiners knew and I knew it too, that I had repeated the fourth year of my studies for having failed the test in seismics. This meant that it took me six years, instead of five, to get an MA degree. Still, my final average mark was much improved, but not enough to improve my job prospects. Here too the system wanted us to choose from a list of available state jobs in the order of the exam league table: the best graduate having 40 options to choose from, whilst the last one was left with Hobson's choice. That was not good enough and certainly not too appealing for my future plans. To avoid being labelled a 'social parasite' all graduates had to accept a job, regardless of where they came in the queue, so I chose a job as a mining engineer. Within a few months of this choice, and following the compulsory military service, I managed a transfer to a publishing house in Bucharest, where a combined knowledge of science and languages was in demand. I was now an employee of the Romanian Academy's Publishing House and I could not be happier for it. Its head was Alexander Graur, a distinguished linguist, whose articles and radio shows I much admired. It was 1967 and I had not given up hope of doing a doctorate in geophysics at some point in the future and in the meantime I carried on my professional correspondence with my academic friends abroad.

NATO SECRET

In the autumn of 1967 I was told by the Newcastle University School of Physics of a forthcoming palaeomagnetism conference at which I was invited to present the results of my Romanian dissertation. This came out of the blue and I was elated.

I was determined this time to wrench a passport from the authorities and fired on all cylinders. I told Newcastle that I would not be allowed out of Romania unless they accepted responsibility for my expenses in Britain. I was later to discover that this was a prerequisite not just of the Communist bureaucracy, but ironically, also of the British Home Office.

"What a splendid convergence of minds", I would have thought years later, but at the time I was just puzzled.

Newcastle sent me the application forms for a travel grant, but very much to my dismay, I discovered that the word 'NATO' was printed on the headed paper. I told them that I stood no chance of obtaining a passport if it had a NATO sponsorship: obligingly, all further correspondence from Newcastle carefully deleted all reference to NATO.

Soon I was sent the conference programme, which included my contribution as well as a return train ticket from Bucharest to Newcastle. This was close to a miracle, because no Romanian citizen could buy an international travel ticket without producing a valid passport, which was impossible to get.

Armed with the invitation, the ticket and the conference programme, I made a fresh passport application. I knew that as a matter of strategy the authorities would not give a negative answer until it was too late: as the deadline for the conference would have passed there was no need for the applicant to go at all! It stood to reason, therefore, to press the Romanian passport authorities for an early answer, so that in the event of the inevitable refusal I could still appeal in good time to attend the meeting.

PICADOR IN ACTION

My friends in Bucharest proved invaluable in affecting the outcome of the application: my English teacher, Madame Jeannette Ulvinianu, had once taught Ceauşescu himself, as well as scores of Government ministers. I also remembered that my girlfriend's mother knew the secretary of the official in the passport office. With the help of these contacts I managed to get some informal interviews, during which I explained that it was an 'honour to represent my country at such a scientific venue' and that I 'had no need for any foreign currency'. I waved the train ticket to Newcastle in front of their eyes—an extraordinary possession, denied to a Romanian citizen without a passport. On appeal, my letter was skilfully placed on top of the in-tray and with a smile and a wink I was invited to get my passport.

The British visa had to be hurried and I did not understand why the British Consul in Bucharest was not best pleased to process the application so quickly. She was truly humourless, in contrast to my high spirits

BALLERINA UPSET

With the British visa granted and duly endorsed in my new crisp passport of the 'Socialist Republic of Romania', I soon realized that, if I were to take the train, the conference would be over. I had to take the plane instead, on a day that 'TAROM', the Romanian airlines, did not fly to London. I would have to change planes in Zurich and from then on travel on a western airline to London and Newcastle, which meant asking for foreign currency, which would have been automatically refused.

The Romanian Airlines office in *Piaţa Universităţii* was packed with disgruntled passengers, each of them with some problem. The official in charge, whom I was supposed to see for a signature, was just finishing an angry exchange with a ballerina from the Bucharest Opera House, whose colleagues were left stranded in West Berlin after one of them had defected:

"Very grave problem, comrade: why did you allow it to happen?" he thundered. I wondered who he was to be concerned about such questions, as I felt it was none of his business to ask. But then, one found Securitate people, or their stooges, in all these places all the time.

When my turn came, the adrenaline still pumping, the comrade asked dismissively what it was I wanted? My chances of success were very slim indeed, so I quipped, in defiance:

"You will not grant me my ticket, because you are already in a bad mood."

To prove the contrary, he looked at the papers and signed the approval.

FIVE GUINEAS GRANT

It was 2 pm on Thursday 4 April 1968. The one-week NATO Summer School conference in Newcastle had started the previous Monday and there was no way I could make it in time. I decided that I would go, visit the University, apologize in person for missing the meeting, return the unused train ticket and return to Bucharest. I rushed breathlessly to the *Banca Naţională,* waving my passport and air ticket. It was to no avail, as I had first to obtain a signature from one of the directors of the Bank: I ran round the corner of the neoclassical building to the main entrance and stopped at the porter's lodge.

The porter was deep in telephone conversation. He did not appear to notice me. I coughed discreetly to no avail, shuffled with no luck and so concluded that the best way to catch his attention would be to run up the marble staircase provoking him to stop me. He did not respond and I found myself walking on the red, plush carpet of the first floor corridor, lined with impressive mahogany doors:

I knocked at random at one of the doors and the smiling face of a blonde secretary beckoned me in—a far cry from the usual unpleasant world of the common bureaucracy: here we were in the 'Land of the Almighty', where things were done politely and without fuss.

"May I help you?"

"I am invited to present a scientific paper in England and I need some sterling."

"Take this paper and make a request in writing. My boss will sign it."

She helped me with the text, took the request next door and within two minutes returned beaming with the approval.

"Here you are, darling! You are very lucky indeed—you have got five guineas, more than what the entire Romanian football team ever get on a trip to England."

I did not know what 'guineas' were, but it sounded like a lot to me.

I mumbled confusedly: "How could I thank you?" and tried to reach for the document, but she would not give it to me.

"Now", she said in a hush, conspiratorial voice, "Tell me darling, who smuggled you into the building, you must have friends in high places?"

"Why?" I protested, "Simply nobody, I promise, I took the grand staircase, as simply as that!"

She would not accept my explanation, as she waved her finger at me, reproachfully: "I do not believe you, young cock, you must know somebody inside!"

She handed over the paper with a wink. I flew down the stairs, past the forgetful porter and got in just in time, only minutes before the bank's foreign currency desk was shut for the day.

OUT IN 24 HOURS

Now I had in my pocket a Romanian passport, a British visa, five guineas and an air ticket to Newcastle. It was 5 pm on Thursday 4 April 1968.

I rushed home to throw a few belongings in a suitcase and arranged to leave by the first plane the following day. I was out of Romania in less than 24 hours from the moment I got my passport. It was not unknown for passports to be withdrawn at the border if one lingered too long. On Friday 5 April 1968 I landed at Heathrow and was granted a visitor's visa for one month.

The same evening I arrived in Newcastle, the very day the conference had ended. I could not imagine, by the wildest stretch of the imagination, that in my pursuit of science or rather because of it, I would be shut out of the Communist prison for 21 years and would not set foot in Romania again until Ceauşescu was brought down!

CHAPTER 2

NATO SECRET

La liberté existe toujours. Il suffit d'en payer le prix.

(Henri de Montherland,
Carnets, 1930–1944)

BETTER LATE...

It was a Friday evening, 5 April 1968, when I disembarked from the plane at Newcastle airport. The NATO Conference on Palaeomagnetism had finished that very day: the School of Physics, which was host to the meeting, knew that I was to arrive and one of their researchers came to meet me off the plane. He was a tall ginger-haired South African, with a lively face and a good sense of humour. The organizers were less surprised that I was late than the fact that I was allowed to come at all. Still, they had arranged a special venue for me the following week, a sort of 'conference after the conference' which I was very pleased to hear about. In the meantime there was an intervening weekend, with all the staff tired in the aftermath of an arduous week-long meeting. I was not to see anybody other than my South African minder, who was going to look after my social programme on Saturday and Sunday. He took me straight to the guest suite which was ready for me, at the top of the University building housing the School of Physics—a penthouse with a superb view over the city. The building itself, I was delighted to hear, was designed by Sir Basil Spence, whose work I knew about and was anxious to discover at first hand. I was handed the keys to the penthouse and the main entrance to the building and my host bade me goodnight saying that he would call the following day at 11 am to show me round.

SHOE SHOPS

It was all too sudden and exciting: I could not stay idly perched in my penthouse and look at the world below, without actually imbibing the newly found environment.

The Northumbrian capital was rather bleak in early spring and the buildings blackened by smoke, but I would not notice any of this at night: I took myself out and slipped into the High Street.

'What was your first impression as you arrived in this country?' I would be asked years later. Upon which I would instantly conjure up the sights of my first evening stroll at 10 pm, on 5 April 1968, in Newcastle. By contrast to the austerity of Bucharest and the rather prudish upbringing required by the norms of the *morala proletariatului* (the morals of the proletariat), I was plunged into a deep sense of amazement if not of shock. Why so many shoe shops, when one or two would suffice, and why would all these youngsters choose to kiss and cuddle in shoe shop doorways? Every single recess was otherwise furnished with couples engaged in a lesser or greater degree of activity, punctuated by heavy heaving and panting, which the low temperature of a northern climate did not appear to diminish a bit. Why didn't they 'do it' in the more discreet comfort of their home?

Upon these musings, I ended my investigative sightseeing and decided to retire to my bed, as I was quite exhausted by my complicated trials and tribulations.

NEWCASTLE BROWN

The following morning I could really enjoy the vista over the city, with the Northumbrian hills on the horizon. The design of the penthouse was brilliant, with lots of glass, which let in masses of light. A spiral staircase linked the apartment to the Common Room below, which had comfortable chairs in bright colours and a functional modern kitchen, a far, far cry from the drabness of Communist architecture: the heart could sing, the imagination could run wild, the world was at my feet.

My South African ginger-haired giant appeared on the dot of 11 am. As I was anxious to glance at the laboratories in the building below we did so then he took me in his car to the pub in Wrexham. The Newcastle Brown ale of which we will hear more later, was de rigueur, although I was not much used to drink at all in Romania and beer in particular was not my scene. But this, I was told, was not ordinary beer, this was 'brown ale', so ale I got.

"Pint?" I was asked.

Not knowing what a pint was and too embarrassed to admit my ignorance, I said "Yes".

Here I was, struggling one sip at a time with a huge glass of the brown concoction, that did not resemble anything I knew or liked, whilst my giant was ready for his second pint:

"Do you like it?"

"Hmmm" I said. "But like all good things one has to savour it slowly", I answered, by way of an excuse.

Our conversation was interrupted by this middle aged Geordie who came and uttered something incomprehensible. My host handed him a coin and as I was expected to do likewise I explained that I only had one five pound note, so he lent me a similar coin which I gave to the Geordie. I could not work out if this had anything to do with our consumption, so I said nothing. A few minutes later the Geordie came back and said something else, in the same incomprehensible language, which made me have grave doubts about my ability to understand colloquial English:

"Nothing to do with it", I was reassured, "This is a different language, more akin to Viking than to English".

"Still, what did he say?"

"Well, he said you did not win."

"What was it then?"

"A game."

"Ah", I said, thankful that the game had stopped: it had already cost me 2% of my capital.

SUDDEN INVITATION

The following Monday I met with all the staff whose papers I had read in Romania and whose research I so greatly admired. Most prominent amongst them were the Head of the School, Professor Keith Runcorn FRS, Professor Kenneth M Creer and Dr David Collinson.

By the time I arrived in Newcastle, Creer had clocked up a most impressive array of articles on world-wide palaeomagnetic data and in particular on the reconstruction of the continents during the Palaeozoic era, 200–300 million years ago. Runcorn's pioneering work in the 1950s on the geophysical interpretation of palaeomagnetic data from America and Great Britain was now a classic of the literature. Runcorn established that the magnetic North pole had been located at different latitudes in the past. However, the data from America and the data from Europe suggested two different polar wandering paths, only converging in recent times. These palaeomagnetic data had supported the possibility of continental drift.

However, by 1968 Runcorn found the subject a bit 'crowded' and moved on to the more rarefied subject of the 'moon and planets', which was the topic of the conference I was invited to and the source of much-needed funding from NATO, as the Americans were busy arranging their landing on the Moon. Appropriately, the phrase 'planetary physics' in Runcorn's department prefigured his main interest, which was going to preoccupy him for the rest of his life and where he made his international reputation.

The atmosphere of the School of Physics in Newcastle was very buoyant and creative, with Runcorn inspiring all the staff, arranging conferences, obtaining much needed cash and sponsorship, building the most extraordinary laboratories and designing the latest equipment, all of which was supported by a band of enthusiastic postgraduate students. Runcorn was away much of the time on PR exercises and during his long absences Creer and Collinson would run the show.

My presentation took place the same day. If the results may have appeared to be modest they certainly gave our western counterparts an idea about the facilities, or rather the lack of them, in Romania: they wanted to encourage further contacts and I felt on the whole that the paper was well, if indulgently, received.

"You must come back this summer as a Visiting Research Student", said Creer.

"I would love to", I answered, knowing full well that once back in Romania I would have to surrender my passport and the door would be firmly shut behind me.

"No chance", I thought to myself, "How naive you can be!"

Little did I know at the time that this invitation was going to come in useful later.

RICHER THAN THE ROMANIAN FOOTBALL TEAM

Unbeknown to me, the Easter holidays were starting within a day or so and everybody in the School of Physics was leaving.

"What plans have you got?" they asked me. "Have you got any money?"

"Yes", I said proudly, "I have five guineas."

"That won't take you very far".

"Never mind", I said, "I am invited to spend Easter with friends in Northampton, and after that I am expected in Oxford, then I will go back to Romania. However, as I missed your conference, for which you sent me a ticket, I would like to return it".

"Ah", they said, "Don't be silly, you will be better off to cash your ticket at Thomas Cook's, but you had better hurry up, because they close their office for Easter".

Don Tarling, the intrepid researcher on the palaeomagnetism of Gondwanaland, accompanied me to Thomas Cook's, who immediately refunded me £80. This filled me with a sense of some financial security, although I somehow felt, in my self-deprecating manner, that it was ill-deserved. So, overnight, I was the happy owner of more cash than the whole of the Romanian football team.

DURHAM: CHICKEN POX ON THE MENU

Runcorn was full of astute ideas: as the show had to come to an end, why not send the young Roman to his new destination? Surely Martin Bott from Durham would fill in?

"Wouldn't you like to see Durham?" asked Runcorn, but by that time his secretary had already told Bott, before I had time to protest, that

"A Professor from Romania would be coming."

So, I said good-bye to my hosts, thinking that I would not see them again, and I was put with my suitcase on the train to Durham. Dr Martin Bott met me at the station and insisted, very much to my embarrassment, on carrying my suitcase. I protested to no avail, especially embarrassed, as in the stuffy hierarchical Communist academic society such courtesies from a senior to a junior would be unthinkable. Furthermore, I was dumbfounded about being presented as a 'Professor', which I was certainly not, and did not know how to correct the misunderstanding.

I knew Martin Bott from his papers on the interior of the Earth, the deep structure of the continental margins and shelf basins and on the global gravity anomalies generated by the Earth's Mantle. He took me straight to the Geology Department, showed me round the lavish facilities and come lunchtime, he hesitated to ask:

"I would like to invite you home for lunch, but we have chicken pox."

"That is fine with me", I said, "I could eat anything." as I thought that he was apologetic about what his wife had cooked.

"Not at all", he protested, "It is serious, we have chicken pox."

"Never mind", I retorted, feeling rather hungry, "I am not fussy, I eat anything."

As Bott's disquiet increased by the second, I felt, at long last, that I must have got something wrong, so I asked: "What is chicken pox in French, maybe I would understand?"

Sadly, Bott's knowledge of the Gallic language did not extend to such mundane topics and he scuttled in search of a dictionary. As it was a weekend, the library and most offices were locked and it took poor Bott half an hour before he came back, brandishing a pocket dictionary in his hand and saying in desperation:

"It is *la rougéole!*"

"Ah, now I understand, no problem—I had the *rougéole* as a child".

"Good, so now you can come."

We were late, he rung his wife to say that we were on the way.

The young Bott, aged five, had chicken pox all right. We established straight away a great rapport. How kind and hospitable the English were. The Romanians were hospitable by nature as well, but Communism dictated that one had to be careful whom one asked home, in order to avoid denunciation. Besides, we lived in such cramped conditions that we were ashamed to ask anybody home: it was altogether a different world.

I was given overnight accommodation on the student campus overlooking the Norman cathedral and castle. It was too short a stay to admire the splendid architecture which surrounded the campus, especially as very early the following morning I was given a lift to Oxford by Dr Ron Girdler, one of the Lecturers from Durham. His geophysical work took him to the East African rift valleys, where he carried out gravity surveys. On reaching Oxford, Ron said:

"I brought you all this way and it cost you nothing!"

I could not understand. In Romania, on geological field trips, one would occasionally get a lift in a horse-drawn cart from a peasant, to whom one would dispense a copper coin or two, but to Ron I felt I couldn't. I believe I did not know how to thank him enough and he must have been disappointed.

OXFORD: PHYSICS APPLIED TO THE HISTORY OF ART

At Oxford I was lodged at the 'Old Parsonage', an enchanting centuries-old hotel, near the Laboratory of Physics applied to the History of Art and Archaeology. The two people in charge of the Laboratory were Dr Aitken, now FRS, and Ted Hall, now a Professor, the latter also the owner of an electronic company at Littlemore, outside Oxford, which was manufacturing the various instruments used by the lab. The team was also publishing a specialist quarterly, called *Archaeometry*. Ted Hall asked me to lunch, where I met his family. He showed me his collection of Tang pottery and told how he bought them at Christie's, only to discover that some of them were fakes: it was the very method he patented at Oxford, in his lab, thermoluminescence, which enabled him to distinguish the fakes from the genuine pieces.

Dr Martin Aitken had a very affable disposition and everybody in the lab was extremely friendly. I mentioned the idea of establishing a similar outfit in Romania and I was given encouragement.

For me, who felt closer to art than to science, this particular line of research, carried out at Oxford would have been an ideal outlet for my repressed natural inclinations. In Romania I had wanted to become an architect, but this had proved impossible in the adverse political backlash that followed the Hungarian Uprising of 1956.

What I could not get myself to admit was that in the Romania of 1968 the resources for starting such a venture would have been fraught with difficulties caused by professional jealousies, lack of funds, lack of contacts with the west and even politics.

Still, I genuinely hoped on my return to get something off the ground.

Ted Hall asked me what I intended to do in London before I left. I mentioned the British Museum, where I absolutely wanted to see the Elgin marbles.

"Have you got anywhere to stay?" I said "No."

"I can lend you our mews house: we will not be there, so you can use it."

This, as it turned out, was an exquisite abode full of electronic gadgetry, well ahead of its time, with remote controls and encoded entrance. What I found so exhilarating was the unselfish generosity of my English hosts, the boundless confidence I inspired which was a great compliment, making me feel humbled but invigorated at the same time. I could not help making a painful comparison with some of my former professors and lecturers in Romania, who were only too glad to keep one at a distance and put one down. The Communist society, which was supposed to be based on equality, was breeding the most inhumane inequality: worse still, it was breeding selfishness, jealousy and suspicion, incapable of encouraging creative thinking, but more of this later.

WANNA ROMANIAN ICON?

With the batteries recharged after the depressed state I was in before I arrived, I was now ready to make the last arrangements before my departure from England. I had borrowed heavily in Romania in order to purchase my air ticket, so I was anxious to repay my debts. As my English friends lavished their hospitality on me, most of my original £85 remained intact. I went to Tottenham Court Road to buy a Phillips tape recorder, intending to sell it afterwards in Bucharest and recoup my travel expenses. As rock music was forbidden by the Communist régime, there was a flourishing underground industry of copying tapes for which tape recorders were essential: anybody who wanted to have a party had to have a tape recorder. However the price of such technology was quite excessive and well beyond the means of ordinary people, sometimes reaching the equivalent of six months income.

Before I left Bucharest I had managed to obtain permission from the Art Commission to take with me two Romanian icons. During my last three days in London I tried unsuccessfully to sell them.

I first tried Christie's, followed by Sotheby's and Phillips, until finally I went to a Bond Street dealer. I remember distinctly that the young man in the shop knew nothing on the subject, even though he had some Russian icons on sale. He got rather upset by some of my comments and tried to gently end the conversation with the 'exit' expression:

"OK?" meaning Is that all? Have you finished? Are you on the way?

Little understanding such a question I answered: "It is not OK at all", and I went on asking questions.

Essentially, I was 20 years too early: there was no market for Romanian icons, although many of the so-called experts could not read Cyrillic and therefore could not tell if the writing was in Russian or in the old Romanian alphabet. Should I have presented my icons as being Russian I might have sold them, but I was too honest to say differently and too proud in my Romanianness. I ended up seeing Mr Dick Temple, a specialist art dealer and the expert at the British Museum, but to no avail.

I took my icons to Paris and said goodbye to England having the feeling, as I saw the cliffs of Dover fade away across the Channel, that I would never return.

CHAPTER 3

PARIS STUDENT RIOTS

"Every body continues in its state of rest,
or of uniform motion in a right line,
unless it is compelled to change that state
by forces impressed thereon".

(*Sir Isaac Newton:* "Principia Matematica", *First Law of Motion*)

PARIS, AS I THOUGHT SHE WOULD BE

I arrived at the Gare du Nord, on 1 May 1968 with the distinct impression that I had been here before. Everything fell into place, as if it was a home-coming. The Roman family had ties with France for several generations, with great uncles and uncles studying here, with reminiscences of pre-war visits, with a French aunt by marriage and with many, many contacts and friends, in all walks of life and of all ages.

It was somehow ironic and I had never imagined that I would see London before Paris. London came very much as a surprise to me, as I had no particular preconception: the images I had of London were from books and postcards, like cameos, which corresponded individually with reality, but which could not form a coherent view of the whole. Yes, I knew a lot about London before coming and especially about the history of English architecture, from the classic textbook of Sir Banister Fletcher. However, I did not know what to make of London and in fact it took some length of time to understand it. London grew on me gradually, revealed itself very slowly, until I became very attached to it, in a quite different way from Paris.

By contrast, I had always imagined Paris as it really was, in the minutest detail, from the Dome des Invalides to the bouquinistes, from the Eiffel Tower to the Marché aux Puces, from the Moulin Rouge to the Cimetière Montparnasse: the great axes of the Champs Elysées, passing the Concorde, all the way to the Caroussel and the Louvre, the majestic yet enchanting river Seine, with the bateaux mouche and the elaborately elegant bridges.

I knew that the re-entry visa into Romania would restrict my time in Paris. That was the date in my Romanian passport by which I would have to return to the confines of my country: a stupid and offensive restriction, but which one had to observe if one was to hope, against hope, that one would be allowed to travel abroad again. It was yet another device by a dictatorial government to fetter its citizens when travelling abroad. There was another important factor to consider, other than the visa: as we were not allowed to take out any foreign currency with us, I could not stay abroad indefinitely.

In Paris, as in London, my plans were to cram in as quickly as possible all the tourist attractions and gain an interview with that guru of palaeomagnetism, Professor Thellier, at the Institut de Physique du Globe in Paris.

ARCHAEOMAGNETISM

Although I rushed insatiably all over Paris, to see all the sights I could and more, I had to allow time for my other passion—the quest for palaeomagnetism. Professor Thellier, of the Institut de Physique du Globe, had a chair in Paris and a laboratory just outside, in the suburbs at St Maur. He did some very important pioneering work in Tunisia, in the early 1940s, by calibrating the Earth's palaeomagnetic intensity, through measurements on fired bricks from Carthage.

François Baudelaire, a French friend from IBM who had close contacts with the professor, arranged for me to be received by him in Paris: I presented him with my CV.

I told Professor Thellier that I had always wanted to do a doctorate and asked him if he would be prepared to direct my studies. He asked me about my MA dissertation in Romania and suggested that I come to see him in his laboratories at St Maur the following week.

Within days of my interview, the student unrest in Paris, which the French later euphemistically referred to as '*les événements*' started to unfold. At the beginning I paid no attention, especially as the first skirmishes were on the left bank and I lived and frequented my friends on the other side of the river. I went to St Maur only to see a different sort of man from the previous week: the old professor was a great scientist but no politician. He was an academic in the most old-fashioned sense of the word and by French standards one would describe him as a '*grand Monsieur*', who reminded me of Couve de Murville. Thellier was completely perplexed by the changes which the students, even in his quiet laboratory, wanted to bring about. Why make a copy cat revolution in the peaceable park of St Maur? I remember in particular an aged PhD student who had been preparing his dissertation for the previous seven years and was nowhere near finishing it: he was the most vociferous and the most irksome of all. If Thellier, as a native, did not understand the events unfolding in front of one's eyes, I, as a visitor from Eastern Europe, had even less comprehension and all I could do was to show sympathy and share his bewilderment. Mindful of the purpose of my visit Thellier offered to direct my research towards a doctorate, provided that I found my own finances. French universities, unlike their British counterparts, had no financial independence and no sponsorship or tradition of scholarships. All studentships came via the government, or were awarded by foreign governments, short of being subsidized by the individual student. Quite a number of students took part-time work to keep themselves alive whilst at school and it was not unusual, therefore, to find specimens such as the irksome revolutionary of St Maur spending their lives on a dissertation—a sort of professional student. Such were not my plans. I told the professor that I was honoured and immensely grateful, but that I had no personal subsidies and that I could not expect any from Romania, therefore my acceptance was dependent on success in finding a scholarship from the French Government.

MAY 1968—THE STUDENT RIOTS

With a French PhD on the horizon I was over the moon, thinking that I would make my father proud, who had worked so hard to see me through school, and also and especially continue the Roman family tradition of studies in France.

I proceeded straight away to the Quay d'Orsay, where I had an introduction to enquire about the possibility of being granted a scholarship by the French Government, now I had been offered a place at the Institut de Physique du Globe. Nothing could have been more untimely and more out of place: I remember it was at the time of the Cannes Film Festival, and the news filtering to the Quay d'Orsay suggested that things were out of hand and that the government was losing control for a moment. General de Gaulle was on a state visit to Romania, enjoying immense popularity and having a rapturous reception from Romanians everywhere, whist at home he was being vilified and his cartoons were on posters everywhere, sporting his big nose and the general's cap.

The Paris walls, especially those of the Sorbonne and of the Quartier Latin, were all daubed with provocative graffiti and the École des Beaux Arts was busy mass-producing revolutionary posters: *professeurs, vous nous faites vieillir.*

LE GÉNÉRAL

De Gaulle, like my old academic friend Thellier, must have been equally bewildered. He tried to restore order. He addressed the French people in his poised thundering voice, but if the French did not want to listen, they all reached for their dictionaries, as the General had such a superb command of language that certain of his words were hardly ever used before:

"... *la chienli, la rogne et la grogne*".

"*La chienli c'est lui!*" retorted the students.

The General had to step down.

The situation in the streets of Paris became rather tense. The French Communist Party was taken by surprise at being led from the front by the students: they attempted to invent new slogans and catch words:

'*Travailleurs immigrés tous unis*', for which, as a Romanian, I felt nauseated.

I simply could not understand. The forces of order were in the streets too and they were not gentle. Public service strikes erupted all over Paris. There was no public transport, no refuse collection and in the middle of the boulevards the mountains of garbage were creating joy for the rodents, which left the metro tunnels to have their feast in daylight. Only a smoke grenade would from time to time interrupt their insolence, making them scurry, along with the rest of the pedestrians, to the relative protection of gangways.

The Sorbonne was a battlefield. Students were using the cobblestones of nearby streets as projectiles against the (CRS) riot police and their *panniers à salade*. Before I knew where I stood and could make my exit from a stricken France, the trains also came to a halt: now I was stranded in Paris with no money, nowhere to go and, to make things worse, with my French transit visa and my Romanian re-entry visa expired.

BY THE GRACE OF THE GOVERNOR OF THE BANQUE DE FRANCE
I asked Minou, *mon ange guardien* in Neuilly, whether she could help. She rang her very good friend, the Governor of the Banque de France, to ask what I could do to obtain an extension of my French visa. At a time when the French viewed students with greatest suspicion and foreigners such as Cohn Bendit as a political liability, asking for a visa extension for a young man from a Communist country was asking the impossible. Not so for the Governor of the Bank. He didn't need to be persuaded that I was no revolutionary.

André de Lattre was the son of a distinguished general of the first war, de Lattre de Tassigny, whose name adorned many a French boulevard. He was an institution in his own right, at the Banque de France. He was a representative of that disappearing species of the *vieille France*, a gentleman who received me most courteously in his offices in the Rue des Petits Champs, tucked away behind the Palais Royal. His personal assistant accompanied me to the Préfecture de Police, in l'Île de la Cité, where, on appointment, I had my Romanian passport endorsed for three months by a courteous clerk. The French visa gave me time until the train strike would be over, now all I had to do was to persuade the Romanian Consul in Paris to extend my re-entry visa into Romania.

RE-ENTRY VISA EXPIRED

I proceeded to the Rue de l'Exposition in the 7th arrondissement, near the Eiffel tower, only to be told by a Romanian official at the Embassy that although he understood that my delay was caused by a *force majeure*, I should return to Romania with the expired visa as soon as practicable, as:

"Nothing will happen."

"Well", I said, "If the visa is irrelevant, why do you still require one?"

"Nothing will happen", the man repeated.

Who was he to tell? Why could he not give me at least some paper of explanation? Once I returned, any excuse could be used against me, to deprive me of future opportunities of travelling abroad, by being branded as 'unreliable'. I could not accept this explanation.

The consul said that "Even if one would ask for a new visa it would take ages."

That I was not surprised to hear, knowing the bureaucracy first hand. However I did not give up: surely there most be somebody to impress upon somebody else that the matter was urgent and had to be resolved.

SECURITATE AT UNESCO

Suddenly I remembered, from my student days in Bucharest, when I translated Romanian poetry into French, that one of the more accomplished translators from my native tongue into French was Professor Alain Guillermou, Chair of Romanian at the École de Langues Orientales in Paris. I rung Guillermou and explained who I was. He asked me home where we talked at length about various aspects of literature and people we knew in common. Then I told him my saga. Professor Guillermou had children of my age and knowing Romania he understood my predicament.

"Go to see Valentin Lipatti at UNESCO, he is the Romanian ambassador there, I know him well so I'll ring him straightaway and will ask him to receive you. Surely, he could help."

I had met Valentin Lipatti a few years previously, when he was Professor of French at the University of Bucharest and I had shown him some of my translations of poetry. He would not remember me, I was sure, but as I had met him before and I knew him to be from an 'old family' and a brother of the late pianist Dinu Lipatti, himself an exile in Switzerland, I felt encouraged by Guillermou's suggestion, which I readily accepted.

Guillermou asked Lipatti if he would see 'Monsieur Roman, un jeune homme qui a besoin de vos conseils'. He neither specified that I was Romanian (with my name I could be taken for French), nor the purpose of my visit.

Lipatti agreed to see me and I met him in the great lobby of the UNESCO building. He was sitting on a bench and his secretary directed me to him. She stayed within earshot, although she was neither asked nor required to do so. She had a severe un-engaging demeanour of a rather sinister style, I thought, but I decided not to fret about it and get on with the purpose of my visit. As I spoke impeccable French there was nothing that Lipatti could go by to see that I was Romanian, until I said what I wanted: his face froze and he immediately asked in a cutting voice:

"Do you still remember how to speak Romanian?"

I found the remark gratuitous and unfriendly. Whilst I reverted to Romanian I pretended not to notice the change in atmosphere. I told Lipatti about the aborted visit to the Consulate, that I needed an extension of my visa, which I hoped he could expedite. At the same time I added that Professor Thellier at the Institut de Physique du Globe had offered me facilities to start a doctorate in palaeomagnetism.

"So, you want to stay in France?" he enquired. "Remember that if you chose to take a doctorate in France, this would be a *political option*"—a strange remark to hear from the lips of an academic and an Ambassador to UNESCO. Before I could reflect any further, Lipatti went on the offensive again, this time suspecting some undeclared 'ulterior motives':

"In any case, if you want to stay in the West, what would your prospects be? At best, you will end up a waiter in a restaurant."

As an Ambassador of a people's republic, *inter alia*—of a worker's republic, I thought this was a gem—looking down on the status of the oppressed workers from the top of the ivory towers erected by the Communist Privilegiatura!

Although I was dumbfounded by his cynicism and wanton impertinence, which had no justification, especially since I was sent to him by a distinguished scholar and friend of Romania, I could not help saying to myself: "Well, comrade, I believe I could do better than that!"

However, I could not think aloud, especially as I was going to return to Romania and who knows, either Lipatti himself, or his beastly secretary, or both, might denounce me.

"Come tomorrow to the Embassy", he said, "I will be there and the visa will be waiting for you by 11.30."

I thanked the Ambassador and left, relieved that the meeting produced some positive result. However Lipatti's attitude smacked more of that of a Party hack than that of a civilized individual as suggested by his family name. Many years later on reading General Ion Pacepa's book *The Red Horizons*, I learned that Professor Valentin Lipatti's main *raison d'être* was indeed that of an officer of the dreaded *Securitate*, the Romanian secret police. Pacepa, head of the Securitate under Ceauşescu, eventually defected in the 1970s to the United States, where he still lives today. Pacepa's disclosure of Lipatti's true colours came to me as no surprise. However it was a great pity, as I felt such nausea when listening to Dinu Lipatti's piano pieces due to the unhappy associations, especially that the surviving sibling and Securitate apparatchik probably still received royalties from his late brother's recordings.

EMBASSY APPARATCHIK

At the time of my UNESCO meeting in Paris, of course, I did not know anything about Lipatti's undercover activities; however, I could not trust him one hundred per cent after his reprehensible bouts of arrogance. So, as a precaution, I decided to go to the Romanian embassy accompanied by a French national. I rung Michel Davideau, my dentist friend who lived nearby, who agreed to come with me to the Romanian embassy the following morning. Predictably, Lipatti was not there and neither was his promised visa. Sure enough, I was expected to arrive on my own and an irate official proceeded to shout at me, in Romanian, asking why was I accompanied? I answered, in a matter-of-fact voice, that I was having lunch afterwards with my friend. As I pointed to him, the official apologized, in French, that he had to speak to me in Romanian, as if my native tongue had to be barked in order to be understood. I was told to call the following day and we left, quite relieved that things did not take a worse turn, still very saddened about Lipatti's deception: muddying the reputation of his late brother, I thought.

I did not bother to go to the embassy again. Instead I rang the day after and the day after and again and again, only to be put off on each occasion. This was no longer a deception, it was an outright lie and an abuse of power, taking away from the individual his own rights, in essence, not being welcome in one's own country. Thankfully my French visa gave me some leeway, until a new solution was found. The problems were piling up. I could not abuse my cousin's hospitality so I decided to move elsewhere. Besides, their flat was in the 19th arrondissement, near the Buttes Chaumont, away from my other contacts. Sure, I learned how to cross Paris on foot, from east to west, from Neuilly to the Buttes Chaumont, in four hours, no mean feat, but one not to be repeated twice a day, every day.

SORBONNE BY NIGHT

The *'événements'* were to cement our friendship when unexpectedly Michael, Arlette and myself were shut overnight in the Sorbonne itself, which was besieged by the riot police. It was a terrifying experience for me even though in retrospect quite funny as it gave us a unique insight into the core of the revolutionaries. It all happened as Michael was driving me along the Rive Droite, to take me from his flat to my cousins' flat one evening. As he listened to the local radio news, all of a sudden he got very excited and exclaimed:

"You know, Constantin, this is the first time ever in the history of French broadcasting that reporters transmit live, without censorship, straight from the barricades of the *Quartier Latin.*"

Without consultation, Michael veered across the Seine by the Assemblée Nationale, into *Boulevard St Germain* and headed straight for the Sorbonne:

"Let us see what happens first hand", he said.

We left the car far enough away, in the Rue des Écoles, for the students not to set it on fire and approached the battle scene cautiously, from the side of the riot police. They looked more like Martians in their riot gear, something I had never seen before, yet they did not seem to mind us. We were not the only onlookers: old grannies perched on their balconies would inspect the length of the street. Various concièrges would stick their necks out of doorways, their poodles barking. Far more exciting, though, would be to watch the scene from the other side of the barricade, from the side of the daring students. They looked like passionate Gavroches, with scarves tied over their noses, to avoid the direct effect of the tear gas, as much as to secure some anonymity. We made a cautious détour, through the rue Soufflot into the Rue St Jacques, to find ourselves behind the 'revolutionaries'. They were busy taking out the cobblestones (*les pavés*) which they were catapulting towards the police some distance away. Our eyes started to cry. The atmosphere was electric: a modern style French revolution. All of a sudden the police encircled us. Panic caused the students and onlookers alike to retreat inside the Sorbonne. We had barely entered the courtyard when the massive gates firmly shut behind us. The University building looked like an impregnable fortress: would the police invade the Sorbonne? How dare they? asked the defiant students, certain of the University's long history of autonomy: the Sorbonne was, by ancient tradition, out of bounds for the authorities.

How silly of us to get involved in such a situation! We felt out of place, but nobody seemed to mind. I was petrified at the thought that should the police break down the gates and find a Romanian student within, this might provide them with the much needed propaganda ploy that 'foreigners fomented trouble'. What if they deported me, or made me do time in jail? Then my good Monsieur de Lattre might think that I *was* a revolutionary after all. What a let down! Michael was also a little annoyed: it was not done for a *bourgeois banker* to mess about with students: his Chairman might question his judgement and he might be demoted, or lose his job. Arlette did not seem to mind at all: she was a born revolutionary and she was devouring the scene, she seemed to be in her element. As we could not influence the events from within or without, we decided that we might just as well do some sightseeing in the Sorbonne. The amphitheatres were alive with *ad hoc* meetings, but no particular structured talk, with everybody shouting down everybody else, with catcalls, laughter and free-for-all interjections. Sub-committees seemed to develop all the time.

On reflection, I should not have been surprised by the bouts of leftist hysteria, as only a few years earlier, the same pro-active French Left deprived a Romanian exiled writer from receiving the prestigious Goncourt prize for his fictional book on Ovid's exile in Tomis. This historical account had poignant parallels between the omnipotent rule of Augustus in Imperial Rome and the rule of the Communist dictatorship in Romania.

The Romanian Securitate took care to spread false allegations about the author's alleged 'fascist' past. The target of this insidious misinformation was none other than Vintilã Horia, a Romanian diplomat of the *ancien régime*, now living in Paris. The French press picked up the story and crucified the great man, with a ferocity matched only by that directed, some years later, at General De Gaulle, in 1968. The tactics were effective—both General and writer had to quit.

VIVRE À DROITE

We moved on. Corridors were daubed with graffiti, some more imaginative than others. Marxist symbols abounded: the clinched fist, the hammer and sickle and the picture of Trotsky. All the old order was being questioned and as proof of it, what better form could such denial take than the fall of all barriers and taboos: in the more sombre corners of the Sorbonne the hot-blooded revolutionaries would spend the excess of their hormones.

"Do you see some of these kids?" Michael asked. "They are kids from Neuilly, and from the 16e and other bourgeois areas of Paris. They were on the barricades earlier, but they will soon drive their deux chevaux back to the comfort of their parents homes, will wash, sleep, enjoy the dinner prepared by their loving mothers and tomorrow evening, quite comforted, will start all over again. *C'est vivre à droite et penser à gauche.* It is a slap in the face of their parents."

I did not pretend to understand: for me it was like a sequence from a Passolini film.

By 3 am the riot police lifted the siege. The revolutionaries dispersed. We too could go back to the comforts of our beds. How lucky I was, though, not to be arrested and deported. Or, maybe, my fear was exaggerated, like some paranoiac sequel from my past in the Communist paradise. Wasn't it Communism that they wanted? I was numb and too tired to philosophize. I too went to bed and slept heavily until midday.

Soon I got tired and depressed at all the Gallic agitation: it was mid June and I felt that time was wasting away and that my stay no longer had a purpose. Although I was trapped, I felt I had to do something and, in any event, I could not abuse my friends' hospitality. True, I could accept a long standing invitation to take refuge in the cupboard of the chambre de bonne in Neuilly, but what if I was raped by my angel turned devil? My imagination ran wild. It was time for a rest!

CONSERVATION AT THE LOUVRE

In the quest of something more serious to do I took the initiative of visiting the Louvre Museum laboratories of conservation and restoration, led by Madame Madeleine Hour. I also called on the director in charge of conservation of the historic monuments at the Musée des Monuments Historiques, at Palais de Chaillot. I talked to the latter about the *maladie de la pierre* and how important it was to use quarry stones from the original quarries from which the building was first erected. In Paris, restorers and architects had a difficult problem to face up to: as the original geological seams were either exhausted or built over, the new stone used for restoration would degrade more quickly, through electrolytic reaction, when used in conjunction with the old stone. I moved on to confess to the director of conservation how I nearly failed my Romanian baccalaureate for indulging in 'illicit' extracurricular reading on the subject of the caves of Lascaux, with their extraordinary prehistoric wall paintings, discovered in the 1940s by Abbé Breuil. The Lascaux caves had just been closed to the public, in order to avoid their imminent loss due to a rampant fungus, encouraged by the visitors' breathing and the artificial light.

"Ah", said the director, "You cannot visit these, even the two sons of Madame Madeleine Hour were not allowed entry to the caves since they were shut. Very few people are allowed in, only experts: you must get permission from the district conservation director, Monsieur Saradet, in Périgueux".

I consulted with my young doctor friend, Didier, from the Hôpital Cochin and we proceeded to draft a letter addressed to Monsieur Saradet. I mentioned that I was a visitor from Romania, a graduate in geophysics, interested in physics applied to conservation and archaeology and that I had carried out research in palaeomagnetism and written articles on conservation. I told him about my great love for Lascaux, which nearly cost my baccalaureate, and that I could not really hope to be allowed in, but I just tried in case he thought that my interest was of sufficient merit. Yes, I was lucky: Monsieur Saradet answered asking me to ring him at his office in Périgueux.

"Do come", he said, "But how are you to reach Périgueux, as there is a general strike and trains are not running. Have you got a car?"

I said, "No, but I have a friend who could drive me."

"Bring him along."

Didier was as overjoyed as I was and only too happy to drive all the way from Paris to central France. We made a beeline for Lascaux, without stopping en route to visit anything else before our destination.

TIMELESS LASCAUX

The caves of Lascaux were the object of great controversy and much concern. They were found on the estate of the Montesqieux family, descendants of the great 18th century philosopher, and as such they were owners of the place, or at least they could control the entrance and derive important revenues. With the improvements in public transport, the advent of the car and much tourism, there was a tremendous demand for the public to view the frescoes. By 1955, the caves were 'modernized' to make access easy for high heels, lit by electricity, ventilated to allow the populace to breathe without perspiring. The show had become big business like a sausage machine and the local tradesmen flourished: everybody was happy when suddenly a nasty fungus developed ready to invade the whole grotto and obliterate the unique art treasures. This fungus, called *la maladie verte*, was *Chlorobotys xanthoficaea*, identified by the Institut Pasteur. It was ready to invade the whole grotto and obliterate its unique treasures unless some drastic measures were taken to halt and then to reverse the process. The visits had to stop until such time that a solution was found. By 1968 no decision had been taken as to the course of restoration or conservation, although the pressure from all quarters was immense to reconsider the closure.

The subject was delicate and politically sensitive: it was at that particular time that we were allowed in.

I have a very vivid memory of the visit. We were a party of only five people: a young art historian, training as conservator, who was acting as a guide, my Paris friend and I plus a couple of archaeologists completed the little group. I knew the frescoes from the monographs I had seen in Romania when I was eighteen. I had to persuade myself very hard that it was reality and not a dream, as the place had a tremendous mystique about it, with the three-dimensional frescoes of hunting scenes with buffaloes running with hunters in hot pursuit. The ochre, the yellow, the orange and the black had an extraordinary potency on the retina: our shadows were projecting themselves on the wall of the cave almost as if they were the shadows of the first hominoid artists. My heart was pounding and I felt a severe chest pain under the emotion of the moment, the first time ever I was made aware of the very strong impression that the artistic oeuvre of a prehistoric man could still hold on a contemporary human being. A certain sense of timelessness invaded me and moving furtively from one chamber to another, the scenery kept its mystic atmosphere.

I promised to write a contribution on Lascaux in the *Revista Monumentelor Istorice*, in Bucharest, to alert my fellow countrymen to the complexities faced by the conservation of ancient monuments. At the same time the article would be a way of thanking my hosts for the unique privilege of visiting these art treasures, now out of bounds.

On the way back to Paris we stopped briefly to see the chateaux of Blois, Chenonceaux, Azay-le-Rideau: the coloured postcards I had admired in Romania came suddenly to life.

PROFESSOR, IS YOUR INVITATION STILL VALID?

The post was still on strike: only the telephones were working. It was June and all of a sudden I remembered that Creer, in Newcastle, had asked me if I would not like to go back as a visiting research student: I rung Ken Creer asking if his invitation was still valid. I told him I could not return to Romania because my visa had expired.

"Ring me back the day after tomorrow."

I did.

"You can come for the summer, we will give you free accommodation, research facilities and one pound a day."

I was over the moon: soon I could return to Newcastle and do the real thing—work on palaeomagnetism!

Michael Conolly took me to the British consul in rue d'Aguéssau, off the Faubourg St Honoré. He sponsored my return visa to England: there was some complicity there, after all, we were hardened 'Parisian revolutionaries'. The trains were still on strike. My Romanian icons paid for my plane ticket to Newcastle. Exactly three months from setting foot in Newcastle for the first time, I was, unexpectedly, back again.

It was July 1968.

CHAPTER 4

PET ON ONE POUND A DAY

"All Science is either Physics, or stamp collecting"

(*Lord Rutherford*)

NEWCASTLE ON ONE POUND A DAY

My return to the fold gave an impression of *déjà vu*. This time Dr David Collinson met me at the airport.

"How nice to have you back", said Collinson. "How was Paris?"

"Horrible", I said, "It's only now regaining some normality: just lots of riots and tear gas. Glad to be back in England."

"Welcome!"

I was taken to the same beautiful penthouse, designed by Sir Basil Spence, with panoramic views over the university and the civic centre. This prestigious abode was intended for important visiting academics, rather than research students such as myself. However, Keith Runcorn, head of the School, was a practical man: he realized that it was infinitely cheaper for the School to put the odd visiting Professor in a hotel in town and keep this Romanian in the Penthouse for the whole summer, rather than the other way round. This view, as I was soon to discover, was not very popular amongst the School's academic staff.

My fellow students in Newcastle were a very cheerful, hard-working, hard-drinking, enthusiastic crowd. For me it was relatively easy to blend in, as I was somewhat on 'home ground', having my abode on the top of the School of Physics. The penthouse was only one level above the Common Room. The palaeomagnetic laboratories were on the ground floor and basement, with easy access to the nearby Student's Union. On the dot of 10 pm, a group of postgraduates, which would include Derek Fairhead (now a Professor at Leeds) and Bill Sowerbutts (now teaching at Edinburgh University), would adjourn for a quick pint of Newcastle brown ale, before closing time at 10.30 pm. Afterwards, some of us would return to the lab to do some more work before midnight. It was an extraordinary atmosphere, poles apart from the Department of Geophysics in Romania: the staff–student relationship was informal and inspiring, the students were devoted to their research, the equipment facilities were generous and a great innovative spirit prevailed.

My £1 daily allowance proved ample for my needs, as the accommodation, heating, water and electricity were provided for and I had no expenses other than food. I found the Union refectory sufficient for my needs and soon got used to the steak and kidney pies, the toad in the hole, the Yorkshire pudding and the apple crumble with hot custard sauce. Two shillings and sixpence were spent on lunch and occasionally I would go to town for a treat of real steak and chips for five shillings. The remaining fifteen shillings a day I could save to send food parcels to my parents in Romania. They were quite worried about my well being in a foreign land and tried to galvanize me into rejoining the Romanian fold. I told them that I was happy and should the Bucharest authorities grant my re-entry visa to Romania I would be home by September, when my visiting Research Studentship in Newcastle would expire. The School of Physics applied to the Romanian Embassy in London to acknowledge my student status in England and, in parallel, my poor old father in Bucharest was desperately trying to enlist the help of the Communist bureaucracy. It was an uphill struggle and an unfair one for someone of his age and dignity.

FISH 'N' CHIPS

Newcastle was not just hard work, but hard play as well. My various colleagues tried in turn to acquaint me with the realities of contemporary England, into which I plunged myself with great enthusiasm. I was out and about several times a week and all weekends, discovering the city and its surroundings or going deep into Northumberland.

One of the 'musts' in my educational programme was a visit to a 'fish 'n' chips' shop: I remember distinctly the greasy cod covered in batter and presented in a paper napkin, which would serve both as plate and napkin;

"Where are the knives and forks?" I enquired.

"There are *no* knives and forks!" my mate answered gleefully, happy to hammer another nail into what he must have understood to be my Communist insularity.

I contemplated for several minutes the chunk of oily fish, pondering over means and ways of eating it, then I had to do what I was told, eat it with my fingers, in the best populist tradition and contrary to my upbringing in a Marxist country. The bemused gaze of my young tormentor enhanced the humiliation even further. I could not wipe my fingers on the already soiled paper napkin, so I asked for another napkin:

"What a waste!" snapped back my self-appointed mentor who proceeded to lick his fingers by way of an initiation, although by that time I felt that he put on an act.

"Surely England was more civilized than that!" I repeated to myself.

CLOSE HOUSE

Close House was the Palaeomagnetic Laboratory that the School of Physics had built in the country, away from all geomagnetic interference inherent to the humdrum of cities. The University of Newcastle was left a huge estate by a generous benefactor, in the shape of a Georgian house in its park-like grounds, landscaped by Capability Brown. A Northumbrian man himself, Capability Brown's ideas of landscape came straight from the superb countryside in which he was born. Close House was situated some distance from the main house, unobtrusively tucked away in a glade of birch trees. The building was specially constructed of non-magnetic materials, meant not to interfere with the very fine measurements of the remanent magnetism of rock specimens. Such specially selected materials in this construction included non-magnetic nails, wires, doorknobs, kitchen sinks, pipes etc. The labs were open plan, with adjacent kitchen and bedrooms, intended for the researchers, who stayed over for several days or even weeks. I went on several expeditions to Close House Laboratory. Once I was taught how to manipulate the equipment, I was left to my own devices. Sometimes I was on my own for days on end, with enough tins of food and biscuits, Nescafe and powder milk to suffice basic needs. Such research spells were a magnificent experience of tranquil concentration, which led to the first palaeomagnetic results from the Cheviot Hills in Northumberland.

BEATLE MANIA

During the late sixties London was in full swing and this mood was very much alive and well in Newcastle. A great *joie de vivre* was pervasive, with endless impromptu parties, some of which took place in the Common Room of the School of Physics. Beatle mania was the rage and the top of the pops chart closely watched. After the strictures of the Romanian Marxist régime, I felt a great sense of liberation, which was both invigorating and inspiring. I soon got to recognize the hit songs of the bands in the top of the pops and I danced like mad: so did lecturers, professors and secretaries—we were all 'pals', or rather, shall I say 'pets', to use a Geordie expression.

"How are you pet?"

"Whom do you mean? Me, 'pet'? Forget it—I have an MA from the University of Bucharest!"

"Who cares?"

Still, my fellow geophysicists did not know the Romanian kissing dance: I had to introduce them to this novel Balkan jinx. So I gave a party where the whole School came to be initiated to panpipe music and the handkerchief dance. Anglo-Saxon caution was thrown to the wind and for once we kissed kneeling on the floor, rather than embracing illicitly in shoe shop doorways, as was the custom in town. Runcorn, too, kissed his secretary.

NO LIQUOR ON SUNDAY!

My pal, Nick Gant, with ginger hair and freckles, was the son of a doctor from North London. He had a Morris van with no side windows, 'to pay less tax', and he offered to take me sightseeing at weekends to visit new places, before I returned to Romania in September. One weekend we crossed the border into Scotland and stopped in Edinburgh.

By comparison to Newcastle, where the Georgian buildings were still covered in a coat of black soot from the Industrial revolution, in Edinburgh, by contrast, the stone was mellow, of a pale cream colour. The streets of the Scottish capital were wide, on the scale of Paris boulevards, with huge vistas: Edinburgh had the reassuring air of a European city. It was only when negotiating the hill towards the castle of the medieval town that a certain Anglo-Saxon character showed itself. Here the architecture was very different from that of England and although less elaborate had a charm reminiscent of the castles I imagined in Walter Scott's novels. On a sunny day Edinburgh looked glorious and lively. Being Sunday and quite hot for a Scottish summer, I entered a newsagent cum grocery shop and noticed on the shelves some bottles of beer.

"Ah, how about comparing the Scottish ale to the Newcastle brown?"

No sooner had I reached for a bottle, than I was rudely stopped:

"Sorry sir, you can't buy this!", said a voice in a strong brogue, rolling its 'r' almost to the point of gutturality.

"Why not?" I asked, in my Romanian accent with open vowels, which made it sound almost like Latin.

He must have thought that I was some Italian ice-cream vendor from Edinburgh, so he retorted in a matter-of-fact way: "Because, you see sir, today is Sunday!"

"Ah", I said, "But what has it got to do with not selling beer on a Sunday?"

"This is the law of the land, sir".

So much for that, but my Romanian experience taught me that the rules were there to be circumvented. I would not give up easily, as my boot-legging spirit came to the fore. After all, nobody was witness to our illicit transaction: there was nobody else in the shop, other than the shopkeeper and myself.

"You see?" I said, trying to shame him, "Even in Algeria, where Muslims are not allowed to drink alcohol, if you show your foreign passport, they would sell you a bottle! Surely you can do the same for a foreigner."

To prove my point I flashed my Romanian passport.

"No way!"

As the shopkeeper started to look aggressive, I knew from Walter Scott's novels that these gillies could be fierce, so I made a hasty retreat. I left the shop empty handed, thirsty and frustrated. As the Scots appeared to be such law-abiding citizens, they would make good material in a Communist country, where dogma was never questioned, just followed blindly.

Latin spirit and Communism never meshed together very well: given time Marxism in Romania would soon be watered down. At this wonderful prospect my face lit with a smile, as I proceeded with the rest of my lazy afternoon stroll.

ROMANIAN OR HUNGARIAN?

The School of Physics Library was very well endowed with specialist books and journals. The librarian was very kind, efficient and knowledgeable and I knew her from my days in Romania, as she had sent so many reprints which stimulated my interest in palaeomagnetism. It was an immeasurable satisfaction to immerse myself in using this library, as I had the feeling that, once I returned home, I would not be able to see a single scientific journal from the West for years to come. Back in Bucharest, only the Head of Department had a subscription to a couple of journals, which he would keep under lock and key until they became obsolete. This restrictive practice would convey the Professor a clear advantage over his peers and turned the whole exercise into a power game, which was our scientific Nemesis.

When I had a moment free from my research lab, I went to the main library: this was looked after by a plump, middle aged lady with a kind smile and motherly demeanour.

On learning that I was Romanian she asked: "Tell me, what language do you speak in Romania, Hungarian?" Quite a shocking question, so I had to pinch myself to believe my ears.

I rationalized: "Did she mean to say that we spoke Hungarian as a foreign language or as a national language? Or, did she conversely mean that in Hungary they spoke Romanian, or simply that the Romanian language did not exist?"

She kept on smiling, whilst waiting for my answer, which took ages to materialize.

"No, would you believe it, or not? We speak Romanian in Romania!"

"So", she said, "You have a language of your own!"

Obviously, she must have thought of the poor Scots, across the border from Newcastle, who were relegated to speaking English, but the Romanians were far more resilient than the Scots: they would not give up their language after a few border skirmishes. They fought for two thousand years to preserve their Latin identity, after the Roman withdrawal from Dacia. The invading hoards of Asians (the Goths, the Hungarians, Gaepidae, Avars, Slavs, Bulgarians, Huns, Petchenegues, Coumans, Tartars and finally Turks), they all tried to impose their foreign tongue on the Romanians, but we stubbornly stuck to speaking Latin, keeping our Romance links, as a kind of 'survival kit', a life raft to the civilized West. For most Romanians of Transylvania, Hungarian was the language of Attila the Hun—the 'Scourge of Christianity'. Wasn't Mr Hitler likened to 'the Hun', by the Brits, during the Second World War? Suddenly, it occurred to me that starting a cultural crusade in Newcastle had a great purpose for Romania and an instant appeal to me.

'ROMANIA COMES TO NEWCASTLE' FESTIVAL

This is how the idea of a 'one-man festival' came to my mind: I called it: 'Romania comes to Newcastle'. A ten-day event which 'should clear the air and put the record straight', for everybody to see that we *had* an identity, we *were* on the map and we *did* make a contribution to that much maligned world 'culture'.

I had to start from scratch, but where would I start? Everything had to be done and had to be done quickly at that, as I was going to leave Newcastle within two months. I wrote to several institutions in Romania: the Romanian Orthodox Church, the Museum of Archaeology in Bucharest, the School of Architecture, asking them to send photographs for an exhibition in Newcastle, the Romanian Academy to send books, to my parents to send my collection of slides. Within three weeks I had enough material for my exhibition: books and photographs, films and slides to make presentations. All these had to be advertised and structured in a programme, a venue had to be found. I advertised for help in the Student Union and soon Peter Micklethwaite, an MSc student in Biology came to offer his time and initiative to help organize various events. We went to the head of the School of Architecture and got on loan a screen of several large panels to hold our photographs in the foyer of the Student Union. These panels were strategically placed on the way down to the canteen, where everybody would have a chance to 'bump' into my exhibition, whether they wanted to or not. Ken Hale, a lecturer in landscape architecture, also came to give a hand with mounting the exhibition, which was organized by themes: architecture, archaeology, Orthodox church etc, not forgetting female beauty in Romania from the Palaeolithic (a goddess of fertility) to modern times (my sister Alexandra Veronica, wearing a Romanian blouse, which reminded me of Henri Matisse's *La Blouse Roumaine*). The Lecture Theatre was hired for films and talks with slides. The bookstand adjoined the photo exhibition. A poster made in a hurry with lino engraving blocks and a roneotyped sheet with information about the ten-day long venue completed the effort. *Courier*, the student union rag, covered the event and published a Romanian recipe. I had no time to sit back and measure the success of the venture. If I were to do so by numbers, I was certainly disappointed as the attendance of the lectures was slim and as to the exhibition, why should the brown ale drinking enthusiasts care about Roman remains of the Danube, or the Neolithic beauties? Those who did care were partly inhibited by what they must have mistakenly taken for Communist propaganda, not, in effect, a private venture born out of outrage against ignorance. However, it was not the numbers that mattered in the end, but the point which I made and the impression it made on people that mattered, not only within the University but outside it: the most extraordinary chain reaction occurred with unexpected benefits and satisfaction.

CULTURAL ATTACHÉ

The Romanian Embassy's cultural attaché, Ştefan Năstăsescu, came from London to see the exhibition. A carefully orchestrated meeting of unlikely minds, between Comrade Năstăsescu and W F Mavor, the retired British Army Major who was the School of Physics' Administrative Assistant, caused my request for a Romanian re-entry visa to be expedited. I needed official blessing for the *fait accompli* of studying in England and this was difficult to obtain in retrospect. Here the 'Romania comes to Newcastle' Festival did the trick, even though it was not meant to do it: it was interpreted that I was 'a good boy', who was best patted on the shoulder, rather than given stick.

However, without my dear father's relentless efforts to liaise with the bureaucracy in Bucharest and cajole them the best he could, my papers would not have been in order. It was important not just for myself, but for my family back home as well, to have my status clarified in order to avoid any unpleasantness or retribution.

What also ensued from this encounter with the Embassy Official were endless invitations to come to the Embassy in London to attend official functions. This I was not very keen on, partly because I was not one of the nomenklatura, partly because the experience with the Romanian Embassy in Paris and Lipatti left me with a bitter taste. Furthermore, as a free man, which I thought I was, I would have been entitled to make a choice and *not* go.

TYNE TEES TV

One morning, during the festival, the School Secretary rushed to me terribly excited: "Will you please call the Tyne Tees television studios as they want to talk to you!". There was a certain look in her eyes which I had never noticed before, but I did not assign it any significance at the time.

In Romania we had television, but at home we had no TV set for three good reasons. Firstly, we were too tired trying to survive and preferred to be more selective by going to a specific film or theatre play or concert or reading a book, rather than see the official tosh on the box. Secondly, the programmes on Romanian TV were full of the most boring, relentless and unsophisticated propaganda, of which we were sick, as we had more than our fair share in the newspapers, the radio, at work, not forgetting the slogans in the street. Thirdly, buying a TV set would have cost a bomb, something like six months of my father's salary, when we hardly had enough money to survive on: the whole idea of television was obscene to me.

I did not get too excited about the telephone call from Tyne Tees TV studios. This was not the case for Marion, the Secretary of the School of Physics: she kept badgering me:

"Did you return that call?"

I did, eventually, because from the few TV programmes I watched on the set provided in the penthouse, I had decided that British TV was rather different, to say the least, from that which I knew in Romania. I still was not terribly hooked on watching TV; besides, at the beginning of my stay in England I could hardly follow what they were saying on TV, as they were talking too quickly.

NOT MUCH HAS CHANGED!

The TV studios wanted to give me 'prime time' on the local news in a live programme: would I come to be interviewed? Yes I would. I was told to come well in advance of my appearance. I was given tea 'to relax', although I was not nervous at all and I could not understand where they got the idea from—maybe other people who were interviewed had butterflies, not me. The interviewer told me that they heard about my one-man show, and as a result, they took to the streets of Newcastle with a cameraman, to ask the good denizens two *key* questions:

"What do you know of Romania?" and "What is the capital city of Romania?"

"Would I wish to comment on the answers they received?"

"Sure", I said, "No problem!"

So we went into the studio, which was now transmitting 'live' and heard the commentator introduce me and what I was up to, which had prompted his little investigation in the streets of Newcastle. Then we saw the clips of the goodly citizens scratching their Northumbrian heads and saying:

"Capital of Romania? Is it Zagreb? No, Budapest? No? I do not know!"

Quite a few said "Budapest", which tied up remarkably well with the gem of the University librarian (bless her), who thought that we spoke Hungarian in Romania!

"How right she was, see? Everybody says the same, so she must be right!"

I watched in disbelief, until a man who had fought in the Second World War, said: "I do not know much about Romania. All I know is that it is a Communist country and not a free country".

"How sensible of him—he did not know much, but he knew the essentials", I thought

"Now", asked the Interviewer, "Would I please comment about what the last man had just said?"

How could I, the idiot? As I had to return to Romania how could I say that it was not a free country? It would have been suicidal if I agreed.

I had to think quickly:

"Well", I answered, "The first visit to England by a Romanian was that of Petru Cercel, Prince of Moldavia, during the time of Elizabeth I. As I can see, under the reign of Elizabeth II, the knowledge on Romania does not seem to have improved very much".

At this particular point, the television chap who was running the show made a robust sign to the cameras, using his left arm to hit the right arm with a clenched fist and in a split second I was off the air. That was all the air time I was allocated and the interviewer accompanied me back to the little coffee table nearby, where we had had our first discussion over tea:

"Thank you very much, it was good of you to come. Would £25 do for your fees?"

I was numb: I did not know that one was paid for such things. I said "Yes": £25 represented my allowance for nearly one month at Newcastle. For £19 I could buy myself a two-piece suit at Burton's, as my Romanian three-piece was starting to look a bit tired.

Before I was out of the studio, the telephone calls flooded in. Would I please take them? There was this couple who went to Romania on holidays and said how lovely it was—I said I was delighted, I agreed and so it went on very much in this vein, for some time. Rather more exciting sounded this young girl who wanted to see me afterwards.

"What have I done to deserve this attention, all of a sudden?"

More was to come on my return to the School of Physics. Overnight I was somebody else, although I did not think of myself of being different from the day before: what on earth hit them? As I entered the Common Room for coffee, everybody shuffled and was over-friendly, trying to talk to me, even the driest members of the staff (and there were a few in the solid physics section of the School, little wonder that it was called 'solids') started to acknowledge me. Only the day before, they were ignoring me, as some irksome foreigner who was 'squatting' in the penthouse. Now, my continuing stay in these luxurious lodgings started to gain some legitimacy:

"After all, he was interviewed on TV and only came to this country yesterday, so he must be good!"

ROMANIAN FOR BUSINESSMEN

The University of Newcastle Linguistics Laboratory rang the Department; they had seen me on television and they wanted to know if I would agree to record for their phonetic library a course in the Romanian language: they 'had one in Hungarian, but not in Romanian'.

I 'must come to put this omission right'!

Quite!

I was handed over a text of 'Linguaphone' called *French for the Businessman*: would I please adapt it to Romanian and record it in their laboratory? Before I knew how much hard work it would entail, I agreed. These were some twenty or more step-by-step lessons, which I studiously translated and adapted to Romanian in manuscript form, lesson by lesson. As I gradually prepared the new text, I would go to the lab and record it. I was paid 30 shillings per hour and was very grateful for it. I was able to send parcels of food to Romania more often, as my daily income suddenly increased by 150% .

Soon the complete set of lessons, which later became *Romanian for the Businessman*, was finished and I sent the manuscript by post to Father in Bucharest, to have it typed and returned. Father was the most reliable partner in any of my new enterprises: he was as good as his word and soon the typescript came back: this contained, to my great surprise, as many as 200 pages of double spaced manuscript. I looked at the amount of work in complete disbelief, not thinking that I could have been capable of cramming in so much work in the space of less than a month. I could not let the laborious effort end here: I had to publish the book.

I told the Head of the Linguistics Department of my intention and, very much to my surprise, he said that they thought that the copyright belonged to them, so I could not publish the text. He had to consult his lawyer about it. I knew nothing of lawyers. I had quite a few lawyers in my family and a great uncle who was a judge, so I was not impressed, neither did I think for a minute that I was meant to be impressed. Why should my friendly linguist wish to use a lawyer? I told him that so far as I was concerned he paid me for the actual recording session and *not* for editing and translating the work and that if anybody had the copyright, that would be Linguaphone in London. His lawyer agreed to my contention. See how enthusiasm could have got me into trouble? But I thought that the question of spreading the good name of Romania and putting the country on the map was far more important than a lawyerish opinion.

London was some six hours away and eight pounds return by second class carriage from Newcastle. I took my Romanian manuscript under my arm and headed straight for St Martin's Lane where the Linguaphone offices were, near Trafalgar Square. In typical Romanian fashion, I did not think of the precautionary measure of making an appointment and I arrived unannounced, about lunchtime, in the editor's office.

"Whom shall I say it is?" asked the secretary:

"Mr Constantin Roman from the Universities of Newcastle and Bucharest".

Sir John Marling, Bt, was the founder and owner of Linguaphone. I was ushered into his small office, full of publishing paraphernalia. He was a Sandhurst-trained army chap, now retired, running this London business as well as his country estates in Berkshire and Gloucestershire. He spoke fluent French, which made our contact very easy, and had a great interest in European culture and history. He did not seem to mind that we had no previous appointment: as it was lunchtime, we started our business discussions straight away over some Italian food around the corner from his offices. He was quite excited about my sudden appearance and although I suspect now that he had no immediate need for my manuscript, he offered to buy it straight away, for two hundred pounds, on condition that I created an index. He gave me an advance of one hundred pounds, saying that the remainder would be paid on completion of the index. I was bouncing with joy. As I was going to find out later, what proved to be far more valuable than the fees paid by John Marling was not the money, but his unfettered friendship over the years and that of Jorie, his wife. They became stalwart allies and substitute parents in the difficult, if exciting, times to come.

THE ROYAL ASTRONOMICAL SOCIETY

Keith Runcorn appeared to take only a detached interest in his students: as his foreign tours multiplied, the staff who stood in for him began to grumble. But, in spite of his absences, he took care to encourage students to attend the Royal Astronomical Society meetings in London. These meetings were infinitely more illuminating than the formal lectures at the University.

There were always several MSc and PhD students from the School of Physics taking the train to London to attend regular meetings at the Royal Astronomical Society and I often joined the small group. I presume not many of us fully understood the contents of the highly convoluted lectures, but it helped us tremendously to structure our thinking, learn how to present an idea to a wider public, maintain a high level of research. The Society's premises at Burlington House in Piccadilly, next door to the Royal Academy, were originally intended as a London home for Lord George Cavendish. Its elaborately decorated interior and sumptuous staircase were only matched by the high thinking of its scientific ethos. The Piccadilly location was convenient for journeys to other points of interest in London and at that time I never anticipated that, one day, I would make a contribution to its *Geophysical Journal*, published here, but more of this later.

THE ROYAL SOCIETY

Keith Runcorn also encouraged us to attend meetings of the Royal Society, which were more accessible for me to understand. In 1968, volcanoes and earthquakes were topical subjects, as plate tectonics was shaping up and establishing its tenets as a theory which could explain the processes which deformed the Earth's crust. All the big names in geology and geophysics were taking part in those meetings, from Cambridge, Newcastle and other British Universities, but also from research establishments in the United States, France and elsewhere. The speakers included Edward Bullard, famous for 'Bullard's Fit', Vine and Matthews of ocean-floor spreading fame, Newcastle's Runcorn and Creer with palaeomagnetic studies, Xavier Le Pichon, a young and promising French seismologist working at Lamont, the American-educated seismologist Father Augustin Udias Vallina, a Spanish Jesuit from Madrid, now head of the Geophysics Department at the Complutense University, and the volcanologist Haroun Tazieff. The latter's books (*Les cratères en feu* and *Les volcans*) were bestsellers in Romania, where they sold out within hours, like hot cakes.

During one of the breaks of a Royal Society meeting, I approached Tazieff, who was a world authority on volcanoes, the founder of the Royal Belgian Centre for Volcanology and co-founder of the International Institute for Volcanological Research.

I introduced myself as a Romanian geophysicist from Newcastle University, but informed him that it was not about geology that I wanted to talk to him, but about poetry, his father's poetry. Quite an unusual venue for such an admission, I said, then added that I knew and admired his father's translations of Romanian poetry into French and proceeded to recite one of the French translations of the 19th century poet Mihail Eminescu.

Scientists were roaming everywhere around us talking geology and here I was reciting an abstruse 19th century romantic poem. I admired Tazieff's writings and documentaries on volcanology, but admired even more his father's work, a distinguished member of the Royal Belgian Academy, who had rendered this Romanian verse so well in French. The eyes of Tazieff lit instantly and I knew that I had focused his attention:

"You must come to Paris to meet my parents: here is my address and telephone number in l'île St Louis".

I was familiar with this enchanted corner of the Seine behind Nôtre Dame: here had lived Georges Sand, Chopin's greatest love. It was on this very enchanting corner of Paris, full of historic links, that Chopin played for the Czartorisky family, at the Hôtel Lambert.

Tazieff was as good as his word: during my next visit to Paris he arranged to meet me at his parents' home on the outskirts of Paris. His mother was an impetuous Russian lady, whose name was passed to the volcanologist: she was formidable in every respect and larger than life. At the age of 70 she had just come back from riding in the forest nearby. She painted in strongly chromatic touches and her canvasses covered the whole house. She would not stop talking, whilst truly I was far more interested in discussing poetry with her mild mannered husband, a man as self-effacing as he was a truly outstanding academic.

HOME OFFICE INTERVIEW

Travelling to London from Newcastle was not always for fun or science: bureaucracy was catching up with me and the perennial visa problem meant that I had also to see the Home Office in 'Petty' France. Whilst in

Paris, I had gone to the *Préfecture de Police* by appointment of the Governor of the Banque de France: here I had to queue like everybody else. In Romania I used to queue for hours, so I did not mind very much waiting for my turn to come, if anything it was instructive to hear the conversations within earshot. One such occasion occurred in August 1968, after the Russian invasion of Czecho-Slovakia, when hundreds of young Czech students were stranded in England and wanted to ask for political asylum. The media reports were full of horror stories of the invasion and the sympathy of the British public was firmly on the side of the victims of this wanton aggression. In the circumstances I would have expected the Czech visitors seeking asylum in Britain to be treated with the same sympathy as expressed by the press and public. Not so at the Home Office:

"Why do you want to stay in this country?"

"Because I love England."

"How can you say that when you have barely been here two weeks, in a holiday camp."

"Yes, but I love England", the youth kept insisting in a quivering, uncertain voice, yet with a disarming honesty.

"But how can you say that when you hardly speak English?"

"Do you speak German?"

"Yes, I do."

"Have you been to Germany?"

"Yes."

"Then, why don't you seek asylum there?"

Did I hear the dialogue correctly? Was it possible? How was it possible? How dare they be so heartless? It smacked of the callousness of Communist satraps, surely it couldn't happen in England! Yes, this dialogue was unfolding in front of me and I was pained to realize it: to say that it had a sobering effect on my ideals would have been an understatement—it was, in fact, shattering. Thank God, it did not happen to me: I had the University of Newcastle behind me and a British friend sponsoring my visit, so I was all right, but those wretched Czech students were most certainly not all right and nobody was there to come to their rescue. In Romania we commonly used the expression 'lying like a newspaper', because of the shameless propaganda of double standards. Yet here was a blatant case of double standards in showing all the sympathy in the press, whilst behind the scenes, the opposite was practised. I decided that, regardless of the political hue, all bureaucrats were the same: 'not to be trusted, not to be trusted at all!' I repeated to myself in dismay.

TOVARITCH BELOUSSOV

The Romanian Government had amazed the whole world by condemning the Russian invasion of Czecho-Slovakia and refusing to send token troops under the aegis of the Warsaw Pact. I was rather proud of our stance, even though I did not trust the régime and its motives. In a way, I felt I was the beneficiary of the Prague Spring, as the Cold War had somehow receded and the rules about granting travel documents became more relaxed.

I knew Prague well, having been there four times on the way to and from Poland, and I loved the city, which I considered a jewel of European civilization. When my friends in the lab in Newcastle told me about the news headlines and I learned that Prague was under the Russian boot I could not stop repeating loudly:

"The pigs, the pigs..." Yes, it was all to be foreseen and yet we were hoping that it might not happen.

Runcorn, our roving ambassador, had just arranged, after some arduous behind the scenes work, for the University of Newcastle to award an honorary degree to a Soviet scientist, Professor Beloussov of the Soviet Academy of Science. The degree ceremony was scheduled in August 1968, just as the Russian troops marched in to Czecho-Slovakia: what a contretemps and Beloussov had just arrived in Newcastle ready to take his degree.

Runcorn was embarrassed, but had to go through the motions as sponsor, so was the University Chancellor, Professor Bosanquet, who had to confer the degree of Doctor Honoris Causa on Beloussov. The University academic staff were on the warpath and voted with their feet in boycotting the ceremony. Hardly anybody was in sight, so, the resourceful Runcorn had to ask all the students, secretaries and cleaning ladies in the University to fill the public gallery in order to furnish the hall. It was rather like Catherine the Great's Siberian villages, except that on this occasion, instead of bogus villages, we had a bogus audience. All rather low key, with poor Beloussov bearing the brunt of the invasion, but he was rather thick-skinned, so he could take it.

WHEN EAST MEETS WEST

Runcorn's initiative in proposing Beloussov for an honorary degree was a reflection of his political philosophy. He wanted to encourage contacts at all levels between East and West and even prophesied, some 27 years before the fall of the Iron Curtain, that 'East and West will come together in some huge melting pot'. I did not share his views, which I dismissed as politically naive. Rather more cynically, I thought that Runcorn referred instead to a nuclear melting pot, some sort of meltdown, like at Nagasaki.

In Runcorn's scheme, I was a smaller cog than Beloussov, but nevertheless I too was part of this same philosophical drive, this idea of East and West having a meeting of minds. Or maybe the 'Romania comes to Newcastle' festival had its impact on Runcorn to such an extent that he asked me if I wanted to start a PhD in September, once my visitor's status had ended. I answered in the affirmative, without any hesitation, and added that I was delighted, but that I could see some difficulties regarding the financial backing of such a scheme, as the Romanians would not give a single penny towards my studies.

"Never mind", said Runcorn, "Talk to Mavor, our admin chap, he will know what to do", and after this elliptic statement, he left on a new international tour. I went to see Mavor, but he knew nothing of Runcorn's intentions, so I had to wait a few more days until he could catch him on the phone, in Alaska:

"What are your intentions regarding Roman's studies here?"

"Send him to the British Council?"

I did go to see the British Council representative in Newcastle, but nothing could have been more ineffectual than this specimen of minor administrator out of his depth: as I was 'not officially sent to England by the Romanian Government', there was no hope there. How could I possibly be *'persona grata'* with the Communist establishment back home, when my father was *not* a Party apparatchik? Any such suggestion of seeking 'official blessing' from Romania was quite absurd and out of the question: this was the routine intended for the sons of Communist Party hacks and I was not included in this privileged class. I felt that the British Council should have known better! So, I went back to poor Mavor.

My tenure as a visiting research student was soon coming to an end and in the absence of any financial backing I would have had to go.

"Ask Runcorn again, he has just come back for a few days."

"Ah, another idea, there is a 'Wilfred Hall' Scholarship offered to science postgraduate students of Newcastle, let us apply."

I applied and was soon short-listed. In spite of this early hope, I did not get the scholarship, which was keenly contested by many good and deserving candidates, all natives of Northumberland.

"What shall we do? What can one do?"

"Let us have him start his PhD", Runcorn said, "and we shall find money later."

"But in the meantime Roman has to survive."

"Well, carry on giving him one pound a day until we find a solution."

I did not complain, how could I? After all, beggars could not be choosers! I started my research in earnest on rock samples from the Cheviot Hills. Brian Embledon took me to the site and we collected a lot of sample rocks. In the middle of all these arrangements, one detail appeared clear: I could no longer stay as a guest in the penthouse once my status changed to that of research student. Quite clearly this accommodation had to revert to its original purpose, to be available for visiting professors, rather than the School's own graduate students. I had to move into a student hostel in Leazes Terrace, a nice Georgian terrace house, only minutes away from the School. I enjoyed the move very much as I was less isolated: I could share the kitchen facilities with other students and I proceeded, for the first time in my life, to experiment with cooking. Romanian males never entered the kitchen for any reason other than asking their wives or mothers if the dinner was ready. In England I had to do like any other student in digs, cook for myself, and I enjoyed every bit of it, reinventing Romanian recipes which I remembered my mother preparing back home. It was cheaper and better than the stews and pies at the Students Union of which I had started to get a little bit bored.

FAIT ACCOMPLI

Earlier in the summer, I realized that I could enhance my chances of success if I filed more than one application to several universities, in order to improve my statistical odds. Such application procedures were completely new to me, as I found them rather tedious with form filling, updating my CV, lining up references. In the Communist establishment there was no need for referees, as we had instead an effective system of informers. The sole employer in Romania was the State and every citizen had a file with the Personnel Department, which we took with us wherever we went. We had no need to say who we were. The employers knew automatically who we were, where we were coming from, or what was our social and political provenance. There were no scholarships and therefore no need for references to be asked or be given. Places at Universities in Romania were awarded on the basis of tests for which a political and social mark was very much part of the equation: indeed it had the greatest weight. Those very few Romanian students who had the good fortune of being sent to study abroad, were always the sons of nomenklatura and the geographical choice was Moscow. Back home, my former Professors at the University of Bucharest would not even dream of being asked for a reference, as such practice in Romania was not necessary. There was instead in operation a system of portable personal files, like a 'convict file'. We were all, I presume, 'social convicts': still, in England, I had no other option but to ask for a reference. I pondered for a split second—after all, my supervisor in Bucharest took for granted the hospitality he received from his colleagues in the West: how could he be seen again on his next trip to Newcastle if he had not bothered to respond? The answer to Runcorn's request to my Romanian supervisor Liviu Constantinescu, for a reference, took ages to materialize. Doubtless, in the process, the Communist Party apparatchiks had to be consulted before a reference could be sent.

"Yes, I was his student, yes I got my MA degree"—as for encouraging me to do further studies my professor could not condone it, as he 'disapproved of a *fait accompli*, even if it was a very good one'.

In other words he could not approve of my initiative of sailing under my own steam without an official blessing. My supervisor in Bucharest knew and I myself knew all too well, that I could not obtain such official blessing, as my family was not one of the privileged few who formed the inner circle of the Communist Party nomenklaturists. As it happened, my Romanian supervisor's letter was going to serve me well, as an illustration of the obtuse and obstructive system I was coming from: it made a good case for me asking my supervisors in Newcastle to provide me instead with fresh references. In Newcastle I had already made my mark as a serious, no-nonsense researcher, a good organizer and team worker with a good potential. By contrast, what a poor sport Constantinescu was to have produced such a pathetic letter, pandering to the Communist order. Thinking of Runcorn's liberal vision, I decided that many more gallons of water would have to flow down the river Tyne, before East and West might came closer together.

In the midst of these gloomy thoughts, a pleasant surprise came from the unexpected quarter of the Romanian Academy: Sabba Ştefănescu knew me less well than Constantinescu, but he sent a warm recommendation. Good old Sabba! He did not flinch and said what he thought, cautiously but simply:

"Mr Roman did not distinguish himself during his undergraduate studies, but to the surprise of the teaching staff he managed to work out a very interesting and original diploma paper, in Palaeomagnetism, for which he was congratulated by his examiners".

This was a cock-a-snook to the Communist hierarchy, of which Ştefănescu had little to fear.

CANADIAN VISA

Amongst the PhD scholarships advertised on the notice board at the School of Physics was one at Toronto with Professor Tuzo Wilson. I applied and, having gone through the preliminary motions, was advised to make an early application for a Canadian visa. I knew that my Romanian citizenship rather complicated things, but nevertheless I had to try. During one of my regular meetings at the Royal Society in London, I took the opportunity to visit Canada House, which was around the corner in Trafalgar Square.

Tuzo Wilson was one of the best names in geophysics, which made the Canadian route to a PhD very attractive. More to the point, for me Canada also had a romantic ring about it, as at the age of 24, when still a student at the University of Bucharest, I published a review of a travel book on Canada. With the fee from this contribution I bought myself a much-needed second pair of shoes. Now these same shoes were going to take me to the very country I wrote about, like a magic carpet, the magic shoes. This very thought filled me with juvenile expectation, as I entered Canada House in London.

Another link with the country was my translations of Canadian po-etry into Romanian. By a quirk of politics Romania had in 1966 established diplomatic relations with Canada, as she was interested in acquiring a nu-clear reactor. As a result of these new diplomatic links, anything Canadian suddenly became topical and I happened to be the only available 'expert' in the country with some scant knowledge of its literature. My sudden publication represented a major coup, as I was allocated a quarter page for my feature article, with several translations of Canadian poems, accompanied by a review of Canadian verse. This literary weekly was very popular and had many a distinguished contributor. As an 'unknown illustrious' and a scientist at that, it was unheard of to be given such an accolade. The fee received from the article on Canadian poetry paid for my first three-piece suit, with which I was attired in Newcastle, but by far the greatest satisfaction was the secret political mischief and joy I experienced at the publication in Romanian of a particular poem by Jean-Guy Pilon: *L'Etranger d'ici.*

In the Romania of the mid 1960s this poem was political dynamite. It could not have been published had the poem been written by a Romanian, because of the political innuendoes, which were sufficient to send the poet to jail. Personally, I identified myself with the Canadian poet and the country it described as that of Romania. The excuse of presenting this in a Canadian guise was unique: I tried and I succeeded and let the public read between the lines, in a country 'where we could not shout our rage and where one was barely allowed to die'.

Fired by these vivid memories of my Canadian experience, I went in high spirits to the Canadian High Commissioner's visa section. I explained that I needed a visa to be allowed to read geophysics at Toronto University. To this end, I presented Professor Tuzo Wilson's letters, offering a place as a PhD research student. The Canadian girl at the immigration desk must have been in her early twenties, younger than me, but she certainly knew her line. She was interested neither in the letters from the University, nor in the fact that a PhD student must have acquired, in the process, such prerequisites as O-levels, A-levels, BA, MA and so forth. She recited instead chapter and verse from the immigration book:

"You see, we can't allow just *everybody* into Canada: there is too much pressure. We need educated people: what kind of O-levels have you got?"

"Forget my O-levels. I am here to study for a PhD, can't you see from this letter?"

She pushed it aside.

"Fill in this form with all your O-levels and A-levels and attach to it all the original test results."

Of course I did not come to England with my MSc diploma, let alone with my original documents showing O- and A-level results. How absurd. How dare she, this 'nincompoop' insist on such trivia, when she could see perfectly well that I was wanted for bigger and better things!

"Listen", I went on the offensive, "You say that in Canada you allow only educated people, yes?"

"Yes."

"Very well. According to your own constitution you should be perfectly bilingual, speak both national languages, English and French, yes?"

"Yes."

"Now", I asked with a flourish, "Do *you* happen to speak any French?"

"No", she admitted, "I don't."

"All right then, it follows that I should be on the other side of the desk asking you questions to allow you into Canada: I have the right to be there, you don't." I stormed out of the grand buildings, being unable to contain my disgust at the petty bureaucracy.

Out on the pavement, in Trafalgar Square, of course I realized I was the loser; but was I? Did I actually need or want to put up with such crass idiocy, such moronic bureaucracy, which I thought was confined only to Communist countries? Surely, I concluded reluctantly, the realm of 'nimbecils' had no boundaries. There must be some international mafia of stupids, who perch in such jobs all over the world.

ENGLISH PROFICIENCY

I wrote to Tuzo in Toronto, saying that my visit to Canada House in London was not a success, as they wanted proof of my O-levels, which I could not produce. He understood my predicament in the hands of bureaucracy, and wrote a very nice letter, stating that he was going to initiate the request for a visa in Toronto to make things easier. In the meantime, would I please fill in some forms for the University Registrar? Amongst the routine require-ments I noticed that all students from non-English speaking countries had to pass the Oxford and Cambridge English Proficiency test. Why 'profi-ciency'? Wasn't I proficient enough? Hadn't I translated from English? Published in English? Given lectures in English to the WI in Sunderland? Well, maybe, but I had to prove it. No I could not. I would have to pass the blessed test and this meant going through a routine and I hated the routines. Also paying a fee for the test. Besides, as I did not have enough money, I simply could not do it, or better still, could not bring myself to do it. Surely the Geordies in Newcastle understood me perfectly well, even the 'nincompoop' at the immigration desk at Canada House understood me perfectly well when I told her off; that meant that I was proficient in English, why should I prove it? I could not.

I wrote to Professor Tuzo Wilson, saying that, reluctantly I had to abandon the idea of a PhD in Toronto.

HELP FROM CHICAGO

A sense of deep frustration started to get hold of me. I needed some advice, advice from a good friend who cared and who knew something about Academia in the West, but I knew nobody. Whom should I ask? I scanned my memory and for a long time could not think of anybody with close connections with Romania and if possible with my family, somebody in a position to give sensible advice about the direction of my studies for a higher degree in the free world. I remembered, all of a sudden, my father's great friendship with a Romanian literatus, now a professor of history of religions at Chicago. This was Mircea Eliade, who in his youth was my father's contemporary at the Liceul 'Spiru Haret' in Bucharest, where they were together in the same Boy Scout movement. As my paternal grandfather was closely involved with and was a leader of the boy scouts (the *Cohorta Buzau* in pre-war Romania), Mircea often spent time at my family's home in Buzau, where he had many a happy holiday. Later on, as students, Mircea, my father, uncle Victor and a group of friends built a sailing boat, called *'Hai hui'* (The Wanderer) which was launched at Tulcea, on the Danube. This was the beginning of an eventful cruise down the Danube to the Black Sea, where they had been caught in a storm and nearly drowned. This momentous episode of their lives, related so vividly by Eliade in his memoirs, *Les promesses de l'équinoxe* (Paris, Gallimard), further cemented the group's friendship. Soon after, Mircea won a scholarship to read history of religions in India, where he sent regular correspondence to father. However, the two friends grew apart, with Father taking on a profession in industry and Mircea becoming a journalist and writer and getting involved in politics. This caused him to remain as an exile in Paris, where he taught history of religions at the Sorbonne and later on at Chicago. When the Communists came to power in Romania, Father remained with his young family in Bucharest, thinking that the Russian-imposed régime would not last and that the Allies would come to our rescue—a calculated risk which proved terribly wrong and for which he paid dearly. By contrast, Mircea's career abroad evolved to become something of a myth in his home country, although he would be referred to only in a hushed voice, as the conspiracy of silence waged by Communists required one not to mention any of the exiles, even those who made Romania proud. Mircea Eliade fell into this latter category.

I wrote to Professor Eliade in Chicago explaining who I was, saying that I 'felt a little adrift' and asking if he had some advice as to what I should do next. I knew he was a busy man, who travelled a lot and did not put much hope in an immediate answer, or in any answer at all, especially since I conjured up in his mind memories of times past, going back some 40 years. After all why should he care?

He answered promptly, with great sympathy and compassion, and directed me to a Romanian exile contact in London, whom he said would help. This was a gentleman I had never heard of before, as I was not frequenting the Romanian diaspora. His name was Ion Raţiu, a former diplomat at the Romanian Legation in London, when Romania was a Kingdom, not a Socialist Republic. Raţiu was a successful businessman with interests in shipping and offices in Regent Street. I rung and said that Eliade suggested that I should see him and he received me immediately. I told him my story and he smiled; he must have heard it repeated so many times before:

"Ask for political asylum", he said.

"I do not want to. I have family back in Romania and I intend to go back. All I want is to be able to study for a doctorate."

Politics did not enter into it. Not so in other people's minds. Like Lipatti's mind in Paris, it was obvious to Rațiu that I had crossed the divide, the Maginot Line and that I had to make a stand. I could not. He did not help, or rather he felt that he could not assist in any way, just harboured a broad smile, at my youthful inexperience.

A LION IN WONDERLAND

Come September and six months to the day when I set foot in Newcastle, my English visa was soon to expire and I would have to renew it. Once in the Home Office building in Holborn I did not know where to go to, but when I explained that I needed a visa I was pointed in the direction of 'Aliens'. I thought that 'Aliens' were people from other planets. *Maybe I came from another planet.* Actually I felt as if I came from another planet. Apart from this strange foreboding, I did not know exactly what 'alien' meant, except that I was put in that particular category, for what obscure reason I could not tell. I did not feel easy about it, as I thought it a derogatory, or discriminatory term. To make things worse, I did not even know how to pronounce the word 'alien': I thought it was pronounced 'alion' as in 'a lion', so I felt rather feline about it and ready to pounce.

Whilst I was waiting I could not help remembering the scene from only a few weeks past, when Czech teenagers stranded in Britain after the Russian invasion of their country were treated so badly by the Home Office:

"Why do you want to stay over here, in this country? You do not even speak English properly." Another English proficiency test, I thought, this time by the Home Office.

When my number was called a middle-aged man met me. He was, I was told later, 'quite a senior officer'. By this I understood that he was in the military, rather than a high-ranking civil servant, although he was in civvies. Unlike the police at the passport office in Bucharest, who wore uniforms, here at the Passport Office in London the civil servants were called 'officers'. I reckoned because they worked in offices, so they had the right to be in civvies to blend more easily with the rest of the crowd.

The situation was quickly summarized: I had arrived in April for one week's conference and I stayed one month in Britain. I then reappeared in July and obtained a three-month visa as a visiting research student. Now I wanted to study for a PhD.

"Well, well, it sounds as if you like it here, you want to stay."

"No", I protested, "I definitely do not wish to stay, I only want to study for a PhD."

"Yes, you would rather stay indefinitely, wouldn't you?"

"No, I wouldn't!" and so on it went for some time, until I shouted:

"Well, you would like to have in this country 'a lion' (sic—meaning an alien) like myself, but I simply do not intend to settle here, all I want is to take my doctorate and go home."

"We will let you know."

So, I returned to Newcastle empty-handed and we played the waiting game. Poor Mavor had to deal with this 'hot potato' and I eventually got a one-year visa until September 1969.

HELLO, KRUSCHEV!

Winter was coming, so Father had sent me an astrakhan hat: he went to the market in Bucharest and asked a peasant producer selling potatoes: "How much is it?"

"Five lei a kilo", the farmer answered.

"No, I mean your fur hat, how much is it?"

"It is not for sale."

"I give you one hundred lei for it."—This was more than the potatoes the farmer had on sale so he took his hat off and handed it over to Father, who promptly sent it to me in Newcastle, to face the rigours of Northumbrian winter.

I took my new fancy fur hat to London. It was made of lamb skin, a lamb of a specific breed, which was called 'astrakhan', as I presumed it originally came from that particular region. In central Europe it was very fashionable and in Romania it was used for men's hats and jacket linings. At a Sunday market in Petticoat Lane my hat caused quite a stir to say the least: the stallholders all waved at me and yelled:

"Kruschev, Kruschev".

They were right: Kruschev also came to London wearing an astrakhan hat, so, I acknowledged the crowds in Petticoat Lane, waving back in a stately manner, except that I had no limousine and no bodyguards.

A NEW SENSE FOR A 'NUISANCE'?

At the end of my first term at Newcastle I received a huge bill for my student lodgings in Leazes Terrace. What should I do? I was living, somehow, from hand to mouth. The financial arrangements were fine for a short period of time over the summer, but not long term, when I had to see to my other needs in the absence of a family who I could fall back on. I had come to England with only a suitcase and in lodgings I had to buy my own sheets and pillowcases, cutlery, plates and winter clothes. I became very worried at the prospect of not being able to pay the bill, so I mentioned it to David Collinson. He was 'the man on the ground', who was always there to lend a sympathetic ear, in Runcorn's absence and would do something about it as he was a sensible and practical man, who would understand. The School had to face up to its responsibility: they could not possibly, on the one hand, encourage me to stay and, on the other, not see to the financial side of it. True, they tried the British Council, the Royal Society, and the local scholarship, then they split the proceeds of a Demonstrator-ship between several research students including myself, but the latter was more of a stopgap, a palliative rather than a long-term solution. The topic caused some embarrassment, as I put the finger on a rather delicate, but pressing subject. Runcorn had bigger and better things on his mind than the pedestrian question of money for his Romanian student: he thought that by doing nothing the problem would go away. In fairness to him, he must have thought that he had done enough: but this was only a beginning and I had at least three years ahead of me: I had to put the matter on a more solid footing if I were to complete my studies.

"You are a nuisance", I heard Runcorn say, one day: I did not know the meaning of the word, which I had never heard or used before. I thought he meant somehow a 'new sense' in two words. What 'new sense' the old man was talking about, I did not know and said nothing, but carried on being a real nuisance.

WHAT ON EARTH IS PETERHOUSE?

I consulted again with Dr David Collinson, who had encouraged me to ap-ply for scholarships elsewhere and also agreed to act as a referee. Professor Creer also had agreed to provide references, so I regularly scanned the ad-vertisements on the School's notice board, in *Nature* and other journals and newspapers.

Professor Du Bois, from the University of Norman, Oklahoma, was carrying out archaeomagnetic measurements on Mexican artefacts from archaeological sites and I was quite excited about it. I sent him my CV and references, together with a letter asking about the availability of funds.

One day, on the School's notice board, I saw a rather odd invitation for a research scholarship at Peterhouse, Cambridge. I say 'odd' because to date all scholarships were linked with some institution calling itself either University or College, but not Peterhouse, on its own, without the name 'College' or 'University' attached to it. What on earth is Peterhouse? Never mind, I applied for the scholarship, not thinking much, or taking it too seriously.

COMPLETE THAT FORM!

Miss Norman, the Peterhouse Secretary, sent the application forms for the Scholarship: apart from the scientific referees required I noticed also a 'character referee'. I asked Michael Conolly in Paris to write. As a Cam-bridge man, he was quite suitable and he knew me from Romania and Paris *and*, I dare say, the Sorbonne of May 1968. Michael was as good as his word: he wrote on the BOLSA (Bank of London and South America) headed paper a nice character reference and sent me a copy. I did not think that I actually had much chance of getting into Cambridge: it was too elitist and somehow I didn't feel up to it. Cambridge was too remote and beyond my wildest expectations. It was better to plod on with the other applications to Oklahoma and Sydney.

I did not proceed with the rest of the information required by Peterhouse: all they had in hand was my character reference and the fact that I wanted to read for a PhD in the Department of Geophysics, but they had no academic references which were crucial. This was not Miss Norman's idea of getting on with her duty: she kept pestering me with reminders:

"Do you still intend to apply for the scholarship"?

What did she care? After all she knew nothing of me other than what Michael said, but she wanted to 'close the file'. Well, in despair I completed the forms and posted them to Peterhouse, but did not think much of it anymore.

Dan McKenzie, from Cambridge, came to deliver a lecture on plate tectonics to an audience at the School of Physics in Newcastle. He was a shy man, younger than me, with a nervous tick that made him more human and approachable. His talk was fired with enthusiasm: one could see that he liked his subject and he managed to electrify the audience. I did not know much about plate tectonics, which had no currency in Eastern Europe, because science was considered to be an ideological domain. As such it was open to the diktat of a handful of political dinosaurs, such as Beloussov: all the best science achievements had to be initiated in the Soviet Union—they were by definition the *crème de la crème*. As Beloussov did not 'invent' plate tectonics, he was against it, so the new theory was not heard of in Eastern Europe, by simple 'diktat'. Now, for the first time, listening to McKenzie I could see for myself that there were other new ideas around and, far more to the point, these ideas were expressed by researchers of my age and younger. I thought there was no need to become a crusty dinosaur in the West before one could make oneself heard: such was not the case in Romania. I told McKenzie about my application for a scholarship at Peterhouse. He said,

"If you are successful, you could come and work with me." I answered that I knew nothing of the subject.

He said, "It doesn't matter, you will learn".

CAMBRIDGE SHORT LIST

Come the Spring of 1969, several scholarship applications were in the pipeline: in the States, Canada, Australia and England, all of which were going to be decided by Easter. Robert ('Bob') Cotton, a Liberal Senator from Australia and Minister of Aviation, tried to encourage me to settle in Oz:

"Don't stay with the Poms, it's not good for you, come here instead" and he agreed to act as a character referee, even though I am not sure that his politics carried any favours with the Aussie academics. One of them came to Newcastle to interview Brian Embledon and me. I don't think there was much empathy there.

News came from Peterhouse: I was shortlisted and would I please come to be interviewed by Professor Sir Edward Bullard, head of the Department of Geodesy and Geophysics at Cambridge. I knew of Bullard's towering reputation in geophysics from his early dynamo theory for the origin of the Earth's magnetic field, in which the field is generated by a liquid iron core undergoing convection. But most importantly I anticipated with some trepidation meeting the scientist who had given acceptability to continental drift by producing the mathematical computer software for modelling the fit of the continents around the Atlantic (see diagram). The outcome—a best fit between Africa and South America at the 500 fathom (approximately 1000 metre) water-depth—was important evidence for continental drift.

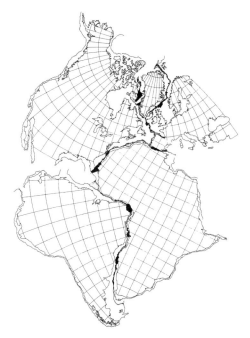

Computer-generated best fit of the continents at the 500 fathom (2000 m) isobath. This contour marks the edges of the underwater continental shelves rather than the coastlines (which change through erosion). Bullard's map demonstrated just how closely the two landmasses fit together and provided important evidence for the theory that the continents had moved. (Reproduced with permission from the Royal Society, from Bullard, Everett and Smith 1965, The fit of the continents around the Atlantic, Philosophical Transactions of the Royal Society of London A **258** *41–51.)*

For this Romanian, the forthcoming interview was like a thunderbolt coming out of the blue: it meant that chances at Cambridge were unexpectedly good, whilst I had put my strong bets on the other horses. I did not think, though, that I would be successful, but I had at least to go through the motions. Allison Clwyd, in Newcastle, had a school friend in Cambridge with whom I could stay. I rang and this was OK, so I was welcome to stay at Richard and Lucy Adrian's home, in Adam's Road.

Lucy's maiden name was Caroe, a family from the West Country, and her grandfather, William Douglas Caroe, was an ecclesiastical architect who designed several bishop's palaces in Canterbury and elsewhere. She was a fellow of Newnham and she lectured at Cambridge in the Department of Geography. Richard was a medical doctor educated at Westminster School and in the States during the war. He too was a typical 'Cambridge product' and stayed on the staff of the Department of Medicine, where he was a lecturer. Richard's father, Lord Adrian, was the recipient of the Nobel Prize for Medicine for his work on neurophysiology. He was a fellow of Trinity College Cambridge and Chancellor of the University. The Adrians were related to whoever was relevant in science, at Cambridge and Oxford, amongst whom were Keynes, Darwin, Bragg, Raverat. To sum it all up, nothing could have been more intrinsically Cambridge than the Adrians, which made my staying with them quite unusual. In spite of their particular status, the Adrians were of easy and uncomplicated manner. Perhaps, in the words of Asquith, a former Liberal Prime Minister, they were members of that inner circle who displayed 'tranquil leadership and effortless superiority'. Nevertheless, I was reminded by Lucy that:

"When the Normans came, the Crockers, the Caroes and the Copplestones were at home", as stated in the age-old verse, to which I added:

"The Romans came to England before that, with Emperor Hadrian."

Richard sensing danger in this game of one-upmanship intervened:

'The Adrians came to England much, much later than anybody else— they were Huguenots'.

A young Bragg nephew, who was an undergraduate at Trinity, was asked to join us for dinner. On learning about the purpose of my visit, he said that he couldn't understand why scholarships were offered to foreigners. I replied that it was a very good thing for this country, for propaganda reasons, as such foreign students were or would become Anglophiles and on returning to their native country they would become leaders and might favour England. He was not convinced of my reasoning. I was rather surprised as to the possibility of any Englishman being anti-foreign, when Britannia 'ruled the waves', had an Empire and the British colonized continents.

IN COMES SIR EDWARD CRISP BULLARD

The following morning, at 10 am, I was expected to be interviewed by Professor Sir Edward Bullard, Head of the Department of Geodesy and Geophysics in Madingley Rise, off Madingley Road.

Sir Edward Bullard's study on the ground floor (bay window) overlooking the croquet lawn and further afield the paddocks and open fields. The Common Room, where many novel ideas in plate tectonics were discussed over coffee and tea, was also on the same floor. Nearby buildings housed the conference room and laboratories, which after Teddy's retirement were called the Bullard Laboratories. (Photo by Constantin Roman 1970.)

The department was located in a Victorian house, complete with its lawns, paddock and gardens. Various laboratories were scattered unobtrusively in the nearby outhouses. It was a romantic setting, with daffodils flowering in profusion and horses grazing in the paddock, which lent the place the mystique of a Constable landscape. Sir Edward had a big smiling face, decorated by a large rounded nose, thin white hair going to baldness. He wore a short-sleeved shirt, with tie, whilst the jacket was hanging on a chair. His informality and constant smile indicated an easy disposition. I did not know what to expect from an interview, so I let him do the talking and ask the questions. We did not stay long in his study, as we soon moved the venue of the interview to the Common Room, where all researchers congregated for coffee. We sat at a long refectory table, near the bay window. There were two other tables in the room, which were soon filled with young people chattering away in a relaxed atmosphere. There was no fixed set pattern and the students mingled with the staff, sitting wherever they chose to. This made for a great opportunity of exchanging news and views. In fact it was on such occasions, at coffee and tea breaks, that some of the most imaginative concepts of plate tectonics were sparked off, with one geophysicist 'bouncing' his ideas off another colleague.

As there were so many people around us, the conversation was not focused and I assumed that they only wanted to have a general idea as to what this Romanian looked like, to ensure that I had no fangs. McKenzie was asked to join us for coffee. I recognized him from his earlier stay in Newcastle and we exchanged pleasantries. We parted and I thought that the interview was over and did not know what to make of it, almost as if it was an anticlimax. McKenzie wanted to know where I was staying in Cambridge and asked me if I was free to join him for dinner in King's. I said that I was staying with friends and that I would let him know if they had any plans for me. On my return to Adam's Road I told Lucy about Dan's invitation to dinner and she agreed that it was important for me and that I should go.

GRILLED AT KING'S

King's, a foundation of Henry VI for the pupils of Eton, was one of the largest and most important of Cambridge colleges. Its importance for English architecture resided in its most resplendent chapel in Gothic perpendicular style, completed under Henry VIII. This was the most perfect example of fan vaulting in England, a feature unique to English architecture. The fan vaulting in King's College chapel was better than that of Windsor or Westminster Abbey. I entered through the neo-Gothic screen in King's Parade, the main college street in Cambridge, and asked the Porter for directions to Dan McKenzie's rooms. I had to take a turn to the left through the 19th century wing of the college, to reach a modern extension where Dan had his rooms. What I found so pleasantly surprising was the lack of inhibition English architects had at combining the contemporary with the old. I could not understand why McKenzie chose to live in a modern block, when he might have enjoyed the Georgian panelling of the Fellows' building across the lawn, but suddenly I realized the motive: Dan had the most pleasant vantage point from his bay window overlooking King's Parade.

I soon came to realize that the object of the invitation was to find out more about my ability and academic background. Romania was a great unknown. We went step by step and explained each of the sixty or so courses I had studied in Bucharest, from geology to geophysics, from physics to maths and combustion engineering. It was more than English students had the opportunity to do in three years, for a BA, and also more comprehensive, in the sense that it covered both geology and geophysics. McKenzie was interested in the Romanian earthquakes and wanted me to work on these. He was definitely counting his chickens before they were hatched, because the scholarship came from Peterhouse, not from the University. We would have to wait and see. He asked where was I staying and I said in Adam's Road.

"Who are you staying with?"

"It's a family called Adrian."

He nearly fell off his chair.

THREE SCHOLARSHIPS IN A GO

On my return to Newcastle I understood that, for lack of money, my stay here had no long-term prospects. Besides, it was getting more and more difficult to solve the accommodation problem and I was increasingly worried about the state of my financial affairs, as the percentage salary from the Demonstratorship was not stretching far enough.

"Ah, you see", said Marion the department's secretary, "You should not have bought yourself a three-piece suit."

I replied, "I brought this with me, from Romania."

I wrote to François Baudelaire, in Paris, to ask if I could not take up a summer job with IBM in Paris. He said that he would let me know as soon as possible. In the meantime, knowing the outcome of the various scholarship applications was just a matter of weeks, I had to play a waiting game, as there was nothing more I could do.

I again explained the situation to David Collinson—and I said that I applied for a student training position with IBM: he understood, so when the positive answer came from Paris I was absolutely elated. There I could wait for the results of my scholarship applications, whilst at the same time I would learn something useful, with a good company and keep myself going.

Sooner than expected, and just before the Easter holidays, the answers from the universities started to filter through, one after another. From the four places I applied for I got three: one in Norman, Oklahoma, one in Toronto (as Tuzo still wanted me to come and kept the option open) and, quite unbelievably, the place at Cambridge.

Only the Aussies did not want me:

"They will be that much poorer for it, damn their desert island and the kangaroos", but I felt, nevertheless, sorry for my old friend, Senator Bob Cotton—somehow I thought that I let him down.

I decided to accept the three-year scholarship from Cambridge and decline the two other places in Canada and the United States, both of which had strings attached with teaching duties. Besides, I started to get tired of all the immigration paperwork, which was complicated by the status of my Romanian citizenship. At least, the British visa was still valid until September 1969, when I could extend it without problems, because of my Peterhouse grant. By contrast, going to Canada or the States would have entailed fresh efforts, starting everything from scratch. I knew that my decision to go to Cambridge was the right one to take and I found the place so fascinating, for its architecture and setting, that I could not think of a more enchanting place on Earth to be.

Now I was ready to go to Paris, to take on my IBM assignment arranged by François, however, there were two or three further details to sort out: my accommodation in Paris, my French visa and my English re-entry visa. Some family friends in Paris, who lived in the 17e arrondissement, near the Arc de Triomphe, learned that I was going to work in the IBM offices at Sablons. This was within walking distance from their flat and offered, free of charge, their servants quarters (chambre de bonne) on the top floor of the building. As a student I had always dreamt of living in an attic room. My re-entry visa in the UK was routinely given, as all my papers were in order and my student status clear. I had only one more thing to do: to obtain the French visa, for four months.

BUT I LIKE FRANCE!

Back again in London, I went to the French Consulate in Tavistock Square and presented my passport. I was immediately informed that

"Romanian nationals had to apply for a visa in their country of origin."

"How very nice and helpful", I thought: "I will go back to Romania and no sooner that I will set foot in the country, the passport will be taken away from me and the door firmly shut behind me. I could then dream of Cambridge and Paris and will not be able to get out of the Communist prison, all this because of the damn French red tape. Bureaucrats of the world unite!" I paraphrased Marx.

I had to do some quick thinking:

"You want me to apply for my French visa in Romania? Sure, but I have to get there first, therefore I would need a transit visa through France", which I knew I was automatically entitled to get on the spot and without waiting. Furthermore, the visa would be ten-days long, which would be sufficient for me to put in position my Banque de France heavy artillery and have my transit visa changed to a visitors visa.

Ah, the French clerk was taken by surprise by my change of strategy. He looked in my passport and noticed that on a previous occasion I had got a transit visa (in May 1968) which I subsequently changed to a short stay, resident's visa. He became suspicious of my nefarious intentions. There were stories of Romanian spies in Paris, of defectors, informers and the like and they could not be too careful. He looked at me with piercing, menacing eyes, full of suspicion:

"Ah, you see, you have been to Paris before, in transit, and you stayed on", he said, full of reproach.

"Yes, I did, because of the train strikes: I could not walk to Romania! I had to take the train and France was on strike."

He did not want to know he had a short memory, especially with regard to such unpleasant details as the *événements*.

"Never mind the strikes", he conceded, "But you *did* stay."

"Yes, I had to regularize my stay and obtain a new visa in Paris. It was the Préfecture de Police which gave it to me, it was all above board."

The clerk changed his line of investigation, for a minute: "But *why*, tell me *why* do you want to go through France? Have you thought of Belgium instead?"

"No", I said, "I have not. Why should I? What is wrong with France? I like France."

"Yes, I can see that", answered the Frenchman, full of innuendoes.

"Now", he said, his face lit for a minute at his new line of attack, "Tell me where shall we send you the visa?"

"Why?" I said, "of course, to the School of Physics, Newcastle upon Tyne." He smiled broadly and handed back my unwanted Romanian passport.

"In that case, mon bon Monsieur, you have to apply for your French transit visa to our Consulate in Birmingham."

"Why Birmingham?"

"Because it is close to Newcastle."

"But I never ever go to Birmingham! I go instead to London, for Royal Society Meetings." He had never heard of the Royal Society, which was irrelevant to the problem.

"But now I am in London", I pleaded, "Can't you do it here"?

"No", he said, "Newcastle depends on Birmingham."

I was quite exhausted by this game of ping-pong and started to lose my temper: I thought it was an abuse of power, of a kind which I knew only in Communist countries. Surely it couldn't be right to be treated like this!

"Well, I would like to see the Consul about that and see what he says". *Eh bien, si c'est comme ça, je voudrais voir Monsieur le Consul.*

"Très bien, Monsieur", he displayed an even broader smirk, "The concièrge will show you the way."

I followed. I soon realized that I was being taken, not to the Consul, but to the front door. I was invited to leave. How humiliating!

"These are no frogs, these are toads, vulgar toads, with no respect for the human race", I muttered to myself in disbelief.

On the train to Newcastle, I thought:

"It was not my fault that I was Romanian and that all these silly restrictions came my way. It was all Yalta's fault, when we were sold down the river by Mr Churchill and the Iron Curtain was put in place." Yet, I had to live with it and somehow circumvent the red tape. At this thought I sighed: what a waste of energy, which could be used for better, more constructive things!

I returned to Newcastle in a dejected mood. I went to Mavor and kindly asked him to do me a last favour and send my passport to the French Consulate in Birmingham. He did. I presume by now he had had enough of writing letters on my behalf, which had became a full time job. He must have been happy in the knowledge that this would be my last fling and that I was going to leave for France. Soon I got my French transit visa, packed my belongings and headed across the Channel to Paris.

It was May 1969, exactly one year, on the dot, after I had first set foot at the Gare du Nord.

A second full circle was now complete.

CHAPTER 5

THE RAT RACE

Chadwick: "You are a lucky man, Rutherford, always on the crest of the wave!"

Rutherford: "Well, I created the wave"

PLATE TECTONICS BANDWAGON

"So, you jumped on the bandwagon", Collinson told me when he learned that I was going to carry out research in plate tectonics. I was interested neither in the politics nor the fashion in Science and I knew very little about plate tectonics. I did not even know what a 'bandwagon' was, as I was unconcerned and totally uninterested in such motives.

I wished I could have carried on with palaeomagnetism at Cambridge, but McKenzie had other plans for me. He was interested in the Carpathian earthquakes and knew that my Romanian scientific background of combined geology and geophysics, doubled by access to native documents, could help make some inroads into the subject. We both accepted that the topic was not sufficiently exhaustive to make the subject of a PhD dissertation, so he added a 'coda', which in effect represented 75% of the workload, *'Seismotectonics of the Carpathians and of Central Asia'*. The 'Carpathian' bit was just for starters, as a means to prove myself as a valid researcher: should I fail at it, after one year, I would be out and I knew it.

Both areas were of evident interest, for different reasons. The Carpathian arc, a mountainous volcanic chain now inactive, was unique because it contained a narrow zone of earthquakes originating beneath the crust. Elsewhere, all sub-crustal earthquakes occurred under subduction zones, where oceanic crust is consumed. In Romania there was no evidence of oceanic crust: just continental crust on either side of the Carpathian arc. The Tethys Ocean had closed during the Miocene, ten million years ago. The closest ocean was the Black Sea, some distance away from the Carpathians, but its crust was defined as being of 'intermediate type' (i.e. a transitional state from oceanic to continental). It seemed, therefore, that there was no way that the Carpathian sub-crustal earthquakes could be caused by the subduction of an oceanic plate within the Eurasian plate.

As for the other subject of investigation, Central Asia (the Tibet and Sinkiang zones behind the Himalayan arc) was a highly seismic area, with epicentres scattered over a wide zone within the Eurasian plate. Again the hyperseismicity of a wide belt of shallow-depth earthquakes within the very body of an otherwise rigid plate contradicted the tenets of plate tectonics, which expected the occurrence of such events along certain lineaments following oceanic trenches, strike–slip faults or oceanic ridges. Central Asia was an exception to the rule—a phenomenon that did not fit the general pattern—which needed further explanation.

By the time I joined the team in Madingley Rise, the handful of American, Canadian and English (read Cambridge) researchers of plate tectonics were busy with the definition of the major pattern of plates: a broad-brush framework which needed refining. By 1969 the reconstruction of the Atlantic opening was already understood and accepted, the details of the Indian Ocean dynamics were soon deciphered and the reconstruction of the western Mediterranean concentrated the researchers efforts. By the same token, nobody as yet had looked at the areas within the Continental plate of Eurasia: from the Carpathians to Tibet and the Himalayas. This was 'virgin' territory, where nobody had trekked before, or rather, nobody had tried to explain using plate tectonic concepts. It was a difficult nut to crack and I was lucky enough to be given the task of finding a solution to this puzzle.

As a research student, I was not obliged to attend lectures, so McKenzie gave me a thick pile of scientific papers on plate tectonics and told me to read them. I was allocated a room in the Department of Geodesy and Geophysics in Madingley Rise, which was at the other side of town from Peterhouse. With money from the first instalment of my scholarship, I bought myself a brand new bicycle, an unnecessary luxury, unheard of amongst most students, who would cycle on improbable contraptions bearing only a distant resemblance to bicycles.

During the first Michaelmas term, I was, owing to my killer instinct, a menace on the Cambridge roads until I learned through my mistakes. I had collided with more than one unsuspecting cyclist and no one could be too prudent in the path of 'the mad Romanian'. My route would take me from Peterhouse to Silver Street, past Queen's and along the Backs, to the University Library, where in the morning, I would pass Lord Adrian, on his bicycle, going in the opposite direction to Trinity. It was good, oxygenating exercise and in the autumn absolutely exhilarating.

VINE AND MATTHEWS SHOW
Coffee was served every morning in the Common Room at Madingley Rise at 11 am and tea in the afternoon at 4 pm. There were no set working hours,

but people would usually come early and leave late, sometimes working well into the evening, as well as Saturdays and Sundays. The coffee and tea breaks in the Common Room gave an excellent framework for keeping in touch with each other, learning about other colleagues' research and exchanging new ideas. It was a relaxed, friendly atmosphere, where the Head of the Department would get a good feel about what went on and supervisors had an extra informal contact with their students. The conversation in the Common Room was scintillating and a propitious occasion for cross-pollinating ideas.

As I was a 'new face' I certainly attracted the attention of this smiling dapper, sun-tanned man, who came beaming at me, with an engaging smile:

"Hi, I'm Drum Matthews, who are you?"

"Ah", I said, correcting his identity, "You are *Vine* and Matthews!"

I had put my foot straight in it. Not a nice way to start the term in the Geophysics Department, but I could not resist it. A killer instinct on a bicycle I had in plenty, but most probably also a self-destructive instinct: how to make things worse, before they got better, in order to succeed in life. However, I was lucky; Drum did not mind at all, or at least he did not show it. Everybody who knew anything about plate tectonics knew about the sea-floor-spreading concept developed by Harry Hess in the early 1960s. In 1963, whilst supervised by Matthews, Fred Vine showed that the oceanic crust on either side of the oceanic ridge was magnetized in alternately normal and reversed remanent polarity, in anomalous bands, parallel to the ridge, as the newly formed crust would magnetize in the prevailing direction of the Earth's magnetic field. They also pointed out what appears to be obvious today, but was unknown at the time, that as the Earth's magnetic field underwent periodic reversals, this would be reflected in the alternate positive and negative anomalies of the oceanic floor. There was an apocryphal story according to which Vine and Matthews sketched their ideas on a scrap of paper in the Common Room, at tea time. It was going to revolutionize the whole of global tectonics and give research a tremendous impetus. This was one of those very simple, elegant solutions that Cambridge was so good at. These observations were of crucial importance in confirming Hess's theory. In 1965, both Hess, from Princeton, and Tuzo Wilson, from Toronto, spent a sabbatical in Cambridge, a cross-fertilization of scientific ideas, which produced papers now regarded as classics of the plate tectonics literature.

By the time I joined the group of researchers at Madingley Rise, in 1969, Teddy had galvanized the success of his Department of Geophysics as a centre of excellence, on a par with the reputation of Rutherford's Cavendish Laboratory in the field of atomic physics: more was to come.

TRANSFORM FAULTS

The new concepts which developed in the Department of Geodesy and Geophysics attracted visiting Professors from Canadian, Japanese and American universities. It was here in Madingley Rise that Tuzo Wilson from Toronto produced his idea of transform faults, which occur when oceanic ridges end abruptly and transform into major faults that slide past each other horizontally. The actual slip between the oceanic plates (deduced from the earthquakes that occur along the fault) is opposite to the apparent displacement of the two ridge sections. This new type of fault offered supporting evidence for the plate tectonics concepts, as the shear movement could be explained by sea floor spreading at the ridge. It was Tuzo Wilson who in 1965 introduced the term plate for the rigid pieces of the Earth's lithosphere.

Tuzo Wilson had offered me a scholarship in Toronto in 1968 and now I had the opportunity of meeting the great man in Cambridge. Here I could talk to the scholar who had made crucial contributions to the theory of plate tectonics: he added support to the sea-floor-spreading hypothesis, by pointing out that the age of islands on either side of mid-oceanic ridges increased with their distance from the ridge. His 'hot-spot' concept had been developed a few years earlier, in 1963. This model explained the formation of active volcanoes, such as those of the Hawaiian islands, in the middle of the Pacific plate, due to the movement of the plate over a stationary 'hot spot'. This idea was considered at first so radical by his fellow geophysicists that it was rejected for publication by all international journals and had to be published in a more obscure journal. When I met him in Cambridge, Tuzo Wilson was at the peak of his career and his theories were vindicated and widely accepted.

In Bullard's Department of Geophysics at Cambridge there were not just 'heavyweights', but also 'featherweights', many from the Massachusetts Institute of Technology, the Colorado School of Mines, the Lamont-Doherty Laboratory, the Scripps Oceanographic Institute and so on. Bullard attracted them all and traffic was two-way, as a lot of our Cambridge staff and students spent their summer in La Jolla and other places, continuing their research. The interaction was close and very productive, the atmosphere was inspiring, almost electric, whilst opening startling new vistas in the realm of geosciences.

ACADEMIC RAT RACE

What was fascinating in this world of 'young gurus' was the mixture of informality and professionalism, of ease in communication and elegance of formulae. Whilst the atmosphere was relaxed, one could sense underneath the tensile drive of working ahead of one's competitors in the field. One of the first tenets of research imparted to me by McKenzie was:

"Publish as quickly as possible."

"Why?"

"Because if you do not do it, somebody else might and pre-empt your work, then all your research will have been in vain."

He did not call this a 'rat race', but in fact it jolly well was. As I was young I did not mind the stress: I plunged myself right into the middle of the unsuspected race and went for it. I aimed at publishing as quickly as possible the first results of my Carpathian earthquakes. After all, my supervisor wanted me to. He was himself a Cambridge product and certainly knew what he was talking about. I had to 'come up with the goods' and I set myself a target of the end of the year. However, as journals usually took a very long time to filter through scientific papers, I had to get the manuscript ready in six months. This in turn meant that I had to aim for the first results within four months, to give myself the time to interpret and integrate the research into text.

STORM IN A *NATURE* CUP

My supervisor was a year younger than I was. This was understandable in view of the fact that in England one could enter University at 18 and after three years get a BA and a further three years a PhD, at the age of 24.

McKenzie obtained his doctorate at 24, after which he stayed on at Cambridge. By the time I joined him in the Department of Geophysics, he had already published several important articles on plate tectonics, defining new lithospheric plates in the Eastern Mediterranean. By contrast, such scientific fulfilment was impossible in Romania, like everywhere else on the Continent, where the situation was rather different: I was a typical case, finishing my MA at 26 (after six years of university, where I read sixty different courses). Now at the 'ripe age' of 28, I was again a postgraduate student, with most of my English peers being around the age of 21. The age difference did not appear to be an impediment and 'Dan', as he preferred to be called, offered enough support and guidance, even though he was rather busy with his own research and was often travelling. This hands-off approach suited my rather independent style of work perfectly, which only needed pointing in the right direction, rather than being tightly supervised. Nobody else in the department worked on seismotectonics and I was Dan's first and only student at that time.

Soon after Michaelmas term 1969, my first computer results, showing the location of the Carpathian earthquakes, had proved well worthwhile. I showed them to Dan and he was quite excited: one could define, for the first time, the shape of the sinking lithosphere under the Carpathian arc of Romania, in the form of a vertical parallelepiped. Now, I had to put some geophysical and geological 'flesh' on these first results in order to give them a plate tectonics meaning and I proceeded to write a first draft. Dan suggested that I should publish the article in *Nature* and so I took care to present the references and figures in the required format. Writing was not as easy as computing the results. Explaining to others what one had done was not immediately apparent, especially when one had to present the whole in a concise, coherent and meaningful manner. I did my best and showed the first draft manuscript to Dan. He returned it soon after that, saying that I should rewrite it, which I did after several weeks. I thought I should show the manuscript to several of my colleagues, for an independent opinion, to see if it made sense. All of them came back with a positive answer, but by then communication with my young supervisor had become more difficult, as he appeared progressively more dismissive. Soon McKenzie disappeared to his American destination for several weeks and I went about my work.

I sent the manuscript for publication in *Nature*, at the end of which I acknowledged my supervisor's help. I thought it important that I should take the risk of publishing and be damned.

I was quite busy in and out of the Department and I did not give my *Nature* manuscript any further thought. I should say I was happy it was out of the way. Within a few weeks the Editor informed me that the two scientific referees were favourable and that the article was accepted for publication, due before the end of the year on 16 December 1970. I was overjoyed, firstly because my angst was unfounded and now I could proceed with my work in confidence; secondly, because this was sufficient proof of my good work to allow me to proceed for a second year and thirdly, it was quite unusual for first-year postgraduates with no established track record, such as myself, to publish in *Nature*, which was the most prestigious scientific journal. Last, but not least, it vindicated my position *vis-à-vis* my supervisor, whom I felt I did proud in his expectation of making me publish 'as fast as possible': it could not have been faster and it could not have been better: I had fulfilled my side of the contract.

FRESH START

New Scientist reviewed my article in *Nature*. This came as an unexpected bonus, which gave further impetus to my work. I needed to keep up the momentum of my research. Sir Edward Bullard was pleased about my promising start and I also shared my joy with Drum Matthews and my younger colleagues, who gave me (metaphorically) a pat on the shoulder:

"Well done, I say, old boy, you've barely arrived and succeeded in publishing in *Nature*?"

I could not fully understand the emphasis which they put on this event and I felt quite modest about it.

Drum Matthews promised to ask Teddy if he would supervise me. The answer was positive. Teddy wanted to get more closely acquainted with my progress: I made a presentation and he seemed satisfied. Now I could address the second and main part of my thesis, the seismotectonics of Central Asia.

I was off to a fresh start.

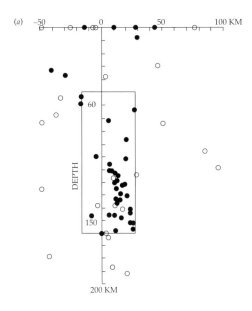

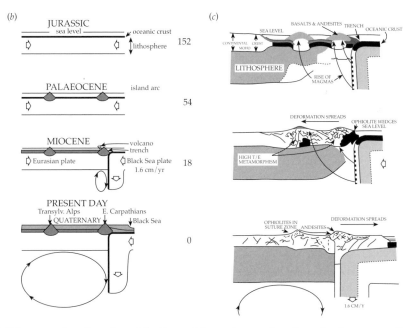

*(a) Presents the relocation of the Carpathian subcrustal earthquakes in cross section with the geometrical definition of the sinking lithosphere under the continental crust of the Carpathian elbow. (b) Intermediate two-dimensional model of the Carpathian plate tectonics evolution. (c) More detailed two-dimensional geological model, which should be viewed in conjunction with the more simplified line model (b). (After C Roman 1970 Proceedings of the Luxembourg European Seismological Meeting ed J-M Van Gils, A, **13** 37–40.)*

IN NEWTON'S SHOES

There could not have been a better outcome than that of working under Sir Edward Bullard's guidance. He had bushy eyebrows, thin, white hair and a constant smile on his face. Forever young, always trying out new concepts, excited with a youthful enthusiasm about new ideas, interested in everything that each of us were doing, always positive and ready to assist. His business and academic contacts attracted the most gifted names in Earth sciences to come to work, or talk to us.

The idea of calling one's boss by his first name had something to do with the way of life in American universities, with whom we had close relations, and perhaps, indirectly, it was also a manner of acknowledging that our Professor, Head of Department, was so young in style:

"Call me Teddy."

Everybody did, except myself, who could not accept such familiarity, which I considered an impiety. Besides, I was marked by the stuffiness of my Romanian professor, whom we addressed as 'Comrade Professor', although he behaved like a dinosaur. It was difficult to shake off this bad Romanian habit, which kept us at an arms length from our professors. By contrast, in England, the relationship between student and staff was relaxed; after all, 'Teddy' could have been our 'father'. He *was* our father, he was at the same time a Professor, a Fellow of the Royal Society, a distinguished scientist of international repute, whose teachers at the Cavendish, starting with Rutherford, stemmed in a direct line from Newton. To me Teddy was Newton's successor: how could I address him as a *'Teddy bear'*? He *was* a 'Teddy bear'!

So, I settled instead for a compromise formula of calling him not 'Sir Edward', which was too formal, but 'Sir E', as his secretary Molly Wisdom called him. Molly was a bespectacled grey haired lady, whose fierce looks and self-righteous opinions were only skin-deep, as she was immensely kind and generous.

Teddy was knighted for his services to science, as Director of the National Physical Laboratory. During the war, when he was only 32, Teddy joined the Admiralty's laboratory of mine warfare on *HMS Vernon*. The Germans were busy sinking British ships at an alarming rate with the newly invented magnetic mines, dropped in shallow waters from aircraft. Teddy understood immediately that the solution to the problem was to reduce the ship's magnetic field, by fitting coils of wire through which an electric current would be passed. Teddy's contribution to developing new techniques for sweeping German mines on the Normandy beaches before landing, is a living example of science's contribution to saving lives and tilting the balance of power in favour of the Allies. After the war, one of Teddy's most significant publications was in 1965 on the best fit of the continents, by closing the Atlantic and making Europe fit North America and Africa fit South America. The results were published in a classic paper with J Everett and A G Smith as co-authors, but as Bullard's name came first in alphabetical

Teddy Bullard and Molly Wisdom in October 1970 at the Blue Boar hotel in Trinity Street, Cambridge. The occasion was the reception following the wedding of Constantin and Roxana at the registry office in Cambridge, where Teddy was a witness. As always, my supervisor was very supportive—the following year he was going to retire to California and I was going to miss him greatly.

order, inevitably the model came to be known as *'Bullard's Fit'*. Teddy once told me that A G Smith bitterly complained to him about this 'misnomer' (that is of crediting only Bullard of all the three authors with the results of joint research): Teddy answered that he 'sympathized with his frustration, but there was nothing he could do about it'.

Under Teddy's leadership as Head of Department, the Department of Geophysics at Cambridge was a world leader in the late sixties and early seventies. It was a privilege to be part of such a team and work under Bullard. It was inspiring and exhilarating.

BANKRUPT SOCIALITE

Teddy came from a Huguenot family which eventually settled in Norwich. Here, Teddy's great grandfather was a publican, who started the brewery of the same name (Bullard's). The family prospered; as a younger son, Teddy's grandfather became an MP. The brewery remained in the family for several generations, until it was taken over by Watney's. Teddy once told me about a beer tasting which they had in the family, at which the Bullard's own brand came bottom:

"It was *very* bad, you know?"

On a journey to Reading University to attend the First European Geophysical Union Meeting, Teddy offered me a lift in his car and we had ample time to converse, which was a great joy. He had the opportunity to reminisce about his family. As he drove along the Thames valley, past Henley, he pointed out to me, high on the river bank, the house of his maternal grandfather, Sir Frank Crisp, from whom Teddy got his middle name. The Crisp grandfather was immensely rich, but got in huge financial difficulties as a result of too lavishly entertaining the Prince of Wales, who was later to become Edward VII.

"Grandfather had to move his overdraft from one bank account to another. His family mansion, *'Friar's Park'*, where the future King was wined and dined with his entourage, is perched on a hill overlooking the river Thames, in Berkshire. Now this home is lived in by one of the Beatles."

Teddy did his research at the Cavendish Laboratory, named after the scientist Henry Cavendish, whose kinsman, the 7th Duke of Devonshire, was the benefactor and founder of the Laboratory. The first Cavendish Professor of Physics was none other than James Clerk Maxwell, whose successor was Lord Rayleigh, followed by J J Thomson (who was only 27 at the time) and Ernest Rutherford. Edward Bullard was Rutherford's pupil. Rutherford's impact on atomic physics was such that in recognition of his achievements, he was buried at Westminster Abbey next to Newton.

The Cavendish Laboratory under Rutherford was a most extraordinary place, where the atom was split and in effect the basis for atomic physics was established. The fundamental research was carried out by a team of young and enthusiastic workers, amongst whom were Blackett, Chadwick, Cockroft, Kapitsa and Bullard to name just a few. These were 'the boys', Rutherford's boys, that is, whose discoveries in atomic physics were to have an enormous impact on the world, especially during the Second World War. It was the 'golden age' of physics at Cambridge, which secured Nobel Prizes for the Cavendish 'boys' with singular frequency: Lord Rayleigh (1903), J J Thomson (1906), Rutherford (1908), W H and W L Bragg (father and son) (1915), F W Aston (1922), C T R Wilson (1927), James Chadwick (1935), George Thomson, son of J J (1937), Blackett (1948), Mott (1977), Kapitsa (1978). Indeed, Cambridge in general and the Cavendish in particular, could boast the largest concentration of Nobel prize winners per square foot. It was common to hobknob with the famous and the powerful, as Rutherford himself recollects:

"As I was standing in the drawing-room at Trinity, a clergyman came in. And I said to him: 'I'm Lord Rutherford.' And he said to me 'I'm the Archbishop of York.' And I don't suppose either of us believed the other."

They most certainly believed in themselves, as Chadwick pointed out to Rutherford about his discoveries:

"You are a lucky man Rutherford, always on the crest of the wave!"

To which Rutherford replied: "Well, I created the wave!"

Rutherford was to inspire Teddy's great ethos, as the latter recalled his former mentor: "He could make you feel a colleague in a great enterprise. His boys were the best boys surrounded by foolish and wicked men". This is exactly how Teddy himself operating in his own Department of Geodesy and Geophysics, some 30 years on in the 1960s and 1970s: we were under his spell—in a way Rutherford's long shadow was still cast in our midst, through Teddy's personality. Or, to put it in other words, we were still riding on the crest of the wave started by Rutherford and his predecessors.

The '*Cavendish tradition*', which was pervasive in Teddy's department, was best summed up by Ritchie Calder, who concluded:

"It broke down the isolationism of research.... it was a case of taking the informality of the common room into the laboratory, retaining and indeed encouraging the individuality of the scientist, however junior and making the professor the mentor, rather than the master of research."

The same struck a Russian visitor to the Department and a contemporary, who wrote to Teddy about his impression of a 'strong and peculiar personal touch, the set of elegant and unpredictable studies, each calling for strong objections (on my part) but altogether revealing a startlingly new geophysical realm'.

After the stultified, Communist-laden, oppressive atmosphere of research behind the Iron Curtain, I felt all of a sudden 'de-pressurized', as I gave free rein to my pent-up enthusiasm, which was exploding. I was lucky, very lucky indeed, to be one of 'Teddy's boys'. Age was no barrier in researching a particular subject. Indeed, in Romania I would have been deemed 'too young' to start a PhD. By contrast, in Cambridge, at the age of 28, I was considered almost a 'mature student'. Nowhere else than in Teddy's department would the poet's verse have been more suitably applied, as '*la valeur n'attend pas le nombre des années*'. Again, in Romania, any such attempts at early scientific fulfilment would have been aborted— gerontocracy ruled OK and it saw that it kept a strong grip, often using ideological means of stifling any budding talent. More often than not researchers would reach the age of wisdom without, as such, reaching the wisdom of the age, like an endemic perpetuation of mediocrity.

I never met Teddy's first wife Margaret, whom I knew had made her own contribution to academia in the form of a book entitled *A Perch in Paradise*. This was inspired by her life as the wife of a don in Cambridge and Toronto and it caused quite a stir, as more than one academic discovered himself under an assumed name. When I went to Heffers bookshop to order the book a few years later, the sales assistant enquired:

"Is it a book about angling?"

Well, maybe—but angling in Cambridge waters was a sport full of surprising and unexpected events, as I was going to find out for myself.

CROCODILE AT THE CAVENDISH

Pyotr Kapitsa, the Russian-born physicist, was Teddy's contemporary at the Cavendish in the 1930s: Teddy told me that Kapitsa would "bathe naked, in the river Cam, during winter, he was quite an eccentric fellow". It was Kapitsa's idea to have the cartoon of a crocodile made in 'sgraffito', by Eric Gill, on the facade of the Mond Laboratory building, which opened in 1933, behind Corpus: *'the Crocodile'* was Rutherford's nickname given to him by Kapitsa. Here the flamboyant Russian émigré (the first foreigner to be elected a FRS, in 1929) worked on the problem of high magnetic fields and liquid helium.

The Crocodile was Rutherford's nickname given to him by Kapitsa. This 'sgraffito' made by the celebrated Eric Gill was fittingly commissioned for the Mond Physics Laboratory in School Lane, inaugurated in 1933. (Photograph by Constantin Roman.)

Kapitsa made his fateful trips to the Soviet Union before the War, the last one of which Stalin aborted, by not allowing him to return to Cambridge. This was in 1934, but Cambridge was generous and sent all his equipment to him in the Soviet Union, a move they came to regret a decade later, as they thought that during the war Kapitsa would work on the atomic bomb. He worked instead on liquid air and oxygen, which was going to boost steel production for the war effort. This work on low-temperature physics brought him the Nobel Prize for Physics in 1978.

It took a good many years, under Kruschev, for the Russian nuclear physicist to be allowed out of Russia again. On his return visit to Cambridge, after some 35 years, Kapitsa would not allow the BBC crew to interview him on television. He acceded to this request only on condition that Teddy would interview him and nobody else. It was during this interview, Teddy told me, that Kapitsa admitted coyly that *'the Crocodile'* was none other than Rutherford:

"In Russia the crocodile is the symbol of the father of the family. But the animal is also regarded with awe and admiration, because it has a stiff neck and never turns its head it just goes straight forward, like science, like Rutherford". By this time, the nickname was an open secret.

Sir E went to Russia on official trips, as a guest of the Soviet Academy of Science. He was also an adviser to the Admiralty and as such attended the Geneva Conference on Disarmament in 1958. Teddy was instrumental in bringing together the first specialists in the field of monitoring underground nuclear tests and he regularly attended Pugwash meetings on the subject. During my research at Cambridge I was going to be an indirect beneficiary of this database in monitoring nuclear tests, when it came to studying the Central Asian earthquakes. During his Soviet trips, Teddy had no qualms that his every movement would be followed, every discussion monitored, his hotel room bugged and bathroom fitted with two-way mirrors. Teddy had an extraordinarily mischievous streak: he told me that he would enjoy, on his trips to Moscow, going to his hotel bathroom and making faces in front of the mirror, to poke fun at his minders, 'in case it was a two-way mirror'.

FUND-RAISER EXTRAORDINAIRE

Teddy was excellent in attracting sponsorship from industry for our projects. This task was made easier by virtue of his position as a Director of IBM and Shell. Money did not seem to be a problem, whether it was a question of going on a scientific marine mission to the Red Sea or the Mediterranean for measurements of the sea floor; fitting a strain meter in Iran to monitor fault movement; spending several weeks at Caltech or MIT; or attending international conferences.

Once, at coffee break, I asked Sir 'E':

"How do you manage to get so much money for our Department?" He answered: "Quite simply, being on their board of directors I tell this or that oil company that it would be good for them if they granted some money to the University, as of course, I could not ask for my own Department. So they say yes, because they gain quite a few tax advantages by giving money away for educational purposes. The University has charity status, you know? So before the grant is due, I go to the Vice Chancellor and say—'look, old boy, I can arrange for the University to obtain quite a bit of money, but you know that we, in the Geophysics Department, need quite a bit ourselves, what with a research vessel in the Red Sea and monitoring earthquakes worldwide': he understands the problem. He knows that most of the grant will have to come to our Department but he has a little left over for his other needs."

"What if he did not agree?" I asked anxiously.

"Well, then he would not get anything at all and he knows it—better something than nothing, so he agrees."

How many of the Cambridge luminaries would have been so good at attracting funds? Certainly very few! In this respect Teddy towered over them all. Although we were aware of the fact that travel grants presented no problem, nobody in the Department abused the system: we only asked if it was necessary and immediately relevant to our research.

I myself could not benefit from this funding largesse in the same way as everybody else did, because my Romanian passport made me less mobile. I always had to seek a visa well in advance of my travels, which somehow complicated the logistics. But I did attend several international seismological conferences in Sicily and Luxembourg. I also travelled to give talks on my research to other universities (Liége, Frankfurt, Oxford, London, Norwich and Newcastle). The Department of Geophysics at Cambridge paid for all the expenses.

Studying the seismological database was relatively easy, because a huge and very accurate file was gathered for the purposes of monitoring atomic tests. The worldwide listening devices, set up by the Americans, were intended to monitor the Russian nuclear tests. To this end, the atomic bomb explosions represented 'signals', whilst the Earth's natural tremors (earthquakes) were 'noise'. For my research it was quite the opposite—the earthquake waves were 'signals' and the atomic explosions were 'noise'. In spite of all of this, or rather because of it, I could be the indirect beneficiary of a sophisticated system, put in place by the Ministry of Defence, at the United Kingdom Atomic Energy Authority (UKAEA), at its Blacknest Laboratory, in Aldermaston. I was equally a beneficiary of the data gathered by the WWSSN, the world-wide seismological station network, established by the United States. The advantage of the latter was that the instruments were identical and therefore the recordings comparable. The data were all on microfiche and one such library was to be found in the Department of Geophysics at the 'Goethe' University, in Frankfurt.

I told Teddy that it was paramount to have access to the WSSN microfiche library in Frankfurt. There was another library in the States, but for me it was too far away and too complicated, because of the visa restrictions on my Romanian passport. Teddy wrote immediately to Professor Hans Berckhemmer, Head of Geophysics at Goethe University, who agreed that I should come over to collect the necessary data on the Tibetan and Chinese earthquakes.

FRANKFURT CHEER

I promised Professor Hans Berckhemmer, Head of the Geophysics Department, to give a seminar on the Carpathian earthquakes and he arranged the venue in the main University Amphitheatre. I tried to structure my seminar in the manner I have seen done in England, that is, starting with a little innocuous joke, in order to keep the audience's interest alive. I had tried to dream up some plate tectonics similarities between Frankfurt and Cambridge and presented some absurd links, which were supposed to be light-hearted. I soon discovered that the audience did not budge a muscle and for a split-second there was a feeling of panic in Berckhemmer's eyes: he must have invited the wrong speaker! I was using phoney language and his prestige was at stake. I had to repair this mistake and soon got the audience under control. Having started my speech in German, merely to apologize for not delivering the whole paper in their native language, went down very well, almost like Kennedy's *'Ich-bin-ein-Berliner'* speech. One sentence sufficed, after which one could even talk in Japanese—the slides did the rest of the job.

I was chuffed to have been given, as a sign of appreciation, a rapturous banging of the wooden desks in the amphitheatre, instead of applause. This, I presume, also happened because it was unusual for a young researcher of my age to address the students, which in Germany was the preserve and privilege of speakers more mature in age and with better credentials.

In the evening, after the seminar, I was invited to Berckhemmer's house for dinner, together with a small group of German seismologists. Here I felt obliged to explain the digression which I made at the beginning of my speech and the purpose of the joke.

"Ah, sooo", exclaimed the professor, "Now I understand" and he proceeded to laugh: it was four hours after I had told the joke. Had he laughed spontaneously, during my presentation, it might have saved me the embarrassment of a failed joke.

"We don't do this sort of thing in Germany. Now I understand why you did it! Ha, ha."

Quite!

PEACE AT ALDERMASTON

Having a lot of raw seismological data on Tibetan earthquakes was great, but it was not enough. My time spent reading the microfiches in Frankfurt proved productive, but now I had to hurry and process this information.

Teddy got in touch, on my behalf, with Dr Hal Thirlaway, a former Cambridge graduate, now the director of the Seismological Laboratory at Blacknest, near Aldermaston, which belonged to the UKAEA. This was a sensitive location, for strategic purposes. I needed clearance, which was given in due course, and I hope that nobody was silly enough to request

Thirlaway to 'keep an eye on me', lest I escape supervision and spirit away some important intelligence for the benefit of the Russians. For me it was another piece in the jigsaw puzzle of my dissertation.

Aldermaston, of course, rang a bell in the recesses of my memory, from when we would watch news reels, in Romania, during the Cold War. The only news on Britain would show Lord Bertrand Russell, squatting in the middle of Trafalgar Square, or his chums marching, from London to Aldermaston on an anti-nuclear protest. I could not imagine how Aldermaston could attract such wrath from the public. 'Fortress Aldermaston'! It did not look like it at all, but the memory was still recent in England and the location still sensitive. That is why the more obscure placename of '*Blacknest*' was conveniently substituted for the Seismological Laboratory, which was monitoring atomic explosions worldwide. At '*Blacknest*', I could use the huge database from the United States Coast and Geodetic Survey (USCGS), which I needed as input for the joint epicentre determination (JED) program for the Carpathian earthquake relocation.

SEISMOLOGICAL CENTRE

The seismological research at Aldermaston could not cover all aspects of my work, so for part of the time I had to go to the International Seismological Centre in Edinburgh, whose director was Dr Patrick Willmore. Pat's generosity at putting at my disposal his Centre's facilities, allowed me to process the focal mechanisms of Central Asian earthquakes. These were analyses of the global distribution of seismic waves, which determined the orientation of the principal axes of compression and tension. Ultimately, this complex computation based on readings of thousands of seismogram microfiches calculated the geometry of the fault which produced a particular earthquake, as well as the slip vectors. The results, in a simplified manner, looked ultimately like many 'footballs' with conjugate compressional and tensional areas. Each little 'football' (mechanism) was the image of the Earth responding to a 'signal' (the earthquake) and showing in stereographic projection zones of compression (black quadrants) and tension (white quadrants). Few people guessed how much painstaking work lay behind these simplified but meaningful sketches.

I was very glad to be in the Scottish capital again, having visited it only briefly, for the first time in the summer of 1968. I loved a lot of things about this city: the topographical position, the architecture, the broad avenues, the fresh breeze of the nearby North Sea, the attitude of people in the street, quite different from southerners. But above all, what I enjoyed most was browsing through the second-hand and antiquarian bookshops in Dundas Street. It is in Edinburgh that I started to buy, for a shilling or two, 18th century French books, beautifully bound in Morocco leather, embossed with gold leaf and Victorian School Prize books, with intricate steel plates. Hardly anybody wanted these, and I could not understand why, but here, at least, they were not sold by the yard, like in Pimlico—a shocking state of affairs, which, I presume, underlined the innate British diffidence towards things 'intellectual'.

"So, you read books?"

Very suspicious indeed.

PLATE TECTONICS ICONOCLASM

With the data now gathered and processed in Frankfurt, Aldermaston and Edinburgh, I was now ready to sit down to interpret it and give it a geological framework.

The crux of the matter resided in the interpretation, in the personalized approach, in the personal touch. The question remained, however—would the results lend themselves to an elegant interpretation? It was not a matter of being 'clever' at all costs—it was rather a question of coming up with a plausible and most of all *original* explanation: this was a *sine-qua-non* condition of obtaining a doctorate—original work.

Now the struggle began, which would separate the wheat from the chaff. I already had some ideas, some intuition, whilst assessing the first pattern of epicentres in Central Asia, and I could start to see certain phenomena more clearly, from the original mumbo-jumbo of epicentres, peppering the whole of the Himalayan hinterland.

Meanwhile, Teddy had become very anxious—the third year was under way and he was pressing me to give a few chapters of my dissertation for his inspection. The first part, concerning the Carpathians, was easily wrapped up and summarized in my *Nature* paper. However, the greater part of the dissertation, the seismotectonics of Central Asia, had yet to crystallize in my mind. I thought the best way to see if my plate tectonics interpretation was 'waterproof', would be to present my ideas to the public, that is the educated public, of British and Continental academia and the international forums of seismology. I was soon to follow an exhausting campaign of 'preaching' the new plate tectonics, as I saw it, warts and all. It was like going into the desert, like the 'early Christian Fathers' preaching to the non-believers. It was a mixture of public relations and striptease, a mixture of defiance and of brinkmanship, which made one terribly vulnerable.

The message, in brief, was the following:

'The plate tectonic concept of rigidity of plates does not work. The whole matter of plate boundaries across continental crust had to be brought into question and the definitions had to be revised.'

Some of the existing plate models were full of contradictions, discrepancies which I had explained in my articles in *New Scientist* and the *Geophysical Journal of the Royal Astronomical Society*. My article in *New Scientist* explained:

"*A basic concept of plate tectonics is the definition of a plate. Usually it is regarded as a piece of lithosphere, comprising oceanic and/or continental crust, delineated by a streamline of earthquakes produced along oceanic ridges and by trenches, major crustal faults and the remnants of oceanic floors, known as ophiolitic belts. Therefore, any such plate marked out by an active seismic belt is not conventionally supposed to contain within its boundaries any earthquakes— it should be seismically rigid, or aseismic. However, such is not the case where the plate boundaries cross continental crust which have been affected by recent Alpine-type mountain-building movements. Over wide areas such as the eastern Mediterranean, Turkey, Persia, Balluchistan, Tibet, Sinkiang and the western United States, there are numerous earthquakes, the occurrence of which would suggest that some of the plates are not rigid and consequently obey different rules."*

Quite understandably, this caused confusion and incredulity in some academic circles, as it had taken some time for scientists to accept plate tectonics and, now that they were accepting its tenets blindly, the rules were brought into question. The facts were there to be seen—the epicentres relocated, the focal mechanisms plotted, the fault plane solutions interpreted, the earthquake magnitudes analysed and correlated to the regional geology. The message came clearly: these are the facts contradicting the theory, this is the new interpretation and my solution—contradict me if you can!

There were no effusive accolades, or cataclysmic demolitions, mostly silent or subdued approval. Geology remains a very conservative profession, where people view change with suspicion. It is nevertheless true that, should the facts have been ambiguous, the old concepts, which I had questioned, would have been fiercely defended. Nobody was defending the barricades—the Roman bulldozer was on its way. In a way I was lucky not to encounter the kind of fierce opposition experienced only ten years earlier by Tuzo Wilson's radical 'hot spot' concept.

LUXEMBOURG

I was soon to find out that Luxembourg had a strong Seismological Institute, which was about to organize a European Meeting on Seismology. My Carpathian study was hot off the press in *Nature* and the idea of plate tectonics applied to this region was completely new.

"What better opportunity for a platform than Luxembourg, to bring the news of my research to the attention of my European colleagues and implicitly to the Romanians?"

As for the latter, they had not been the most helpful to me as a student at the University of Bucharest.

I sent an abstract to the Conference organizers and was soon informed that they had accepted my paper.

On arriving in Luxembourg I unpacked only to discover that, in the haste of leaving Cambridge, I had taken the wrong set of slides for my conference presentation. I was numb with disbelief. What should I do? Cancel my talk?

That would not be very well received and would only give grist to the mill of my Romanian competitors, of whom there were three present in Luxembourg.

"Most certainly not!"

I had a closer look at my slides and decided that, as nobody knew much about plate tectonics, they would not be any the wiser about the discrepancy between the title of my talk and its contents. I went ahead and filled in my allocated twenty minutes.

Did anybody notice?

Nobody did, except myself. In any event, I could send the right illustrations to be included in the conference proceedings and there again, nobody would notice the switch.

'TOO YOUNG' FOR SCIENCE

My presentation in Luxembourg caused a commotion amongst the Romanian seismologists: one of them my former Head of Department, Professor of Gravity and Magnetics and erstwhile supervisor, Professor Liviu Constantinescu. Some two years previously he had written to Runcorn, in Newcastle, saying that he 'could not approve of my staying on in England to take a PhD, for having put the authorities in front of a *fait accompli'*.

The second participant in the Romanian delegation, was my former lecturer in seismology from Bucharest, through whose good offices I had twice failed my exam and was made to repeat a year. He could not understand at all how one of his former students, whom he failed at the exam, could now present a paper with new ideas to an international conference! Had I still been in Romania, these two seismologists would have made absolutely sure that I was never allowed to air any views in public, but as I was in Cambridge, supervised by Bullard, there was precious little that they could do to stop me. The third Romanian participant at the conference was a researcher from the Seismological Institute in Bucharest, who, very much like his bosses, considered the Romanian earthquakes to be his sole and exclusive prerogative, on which nobody else could make a pronouncement.

My model of '*plate tectonics of the Carpathians'* had surprised not only the Romanian team, but also the audience at large. Plate tectonic models had been applied to the Atlantic, and the major plate boundaries were defined worldwide. In Western Europe attention was concentrating on the Iceland Ridge, the Bay of Biscay and the Western Mediterranean, for which several models of reconstruction were published. However, little if any thought was given to the Carpathians, being an area of continental crust, which was supposed to be 'stable' and therefore not part of any active plate boundary.

Furthermore, in Eastern Europe news travelled slowly, whilst in the Soviet Union plate tectonics was completely 'banned', almost by decree.

There, Beloussov, the all-powerful geophysicist of the Soviet Academy, had strong views against it and did not allow anybody else to express an opinion. There was little danger of scientific dissent in Russia, as all scientists were kept in the dark on developments in the West, by simply being starved of the mainstream of information: there were no Western scientific journals circulating in Russia, or very few.

I noticed in the audience my erstwhile supervisor from Bucharest, Professor Liviu Constantinescu, and went to shake his hand. There were no congratulations from him, just surprise that I was 'no longer at Newcastle, rather at Cambridge and researching a different subject'.

"You are a bit young for plate tectonics", he quipped in Romanian.

Although I was not surprised at my old Professor's reaction, I was not reconciled with his attitude, which I thought gratuitous and hurtful:

"Why do this to me? What have I done wrong?"

I had done nothing wrong, other than taking a political option, by being true to my own self—I was free to exercise my profession and research the way I wanted, without any hindrance from the Nomenklatura, without kowtowing to the almighty, without excuses for being alive. Memories came flooding back of past snippets of dialogue:

"May I speak, Comrade Professor?"

"Would you please receive me for five minutes, Comrade Professor?"

Then, of the unspoken word:

"Never mind if you made me wait for hours in your antechamber! This is your little power game, we got used to it."

Well, I did not! I never got used to it! I never accepted that attitude of subservience! Was Liviu Constantinescu still afraid of my lack of subservience, even from such a distance?

SEISMIC ENGINEERING *AL ITALIANA*

My friends in the Geophysics Department at Cambridge were busy arranging their annual summer expeditions to exotic far-flung places: the Red Sea, Aden, Cyprus, the Eastern Mediterranean, Iran and even Eritrea, before all the troubles began. Mindful of both my junior and foreign status, as well as of the restrictions imposed on my travel by my Romanian passport, I did not attempt to join any of these. Besides, the subject of my dissertation would, at best, send me to Romania, which was to be avoided, or to Tibet, which was also in Communist hands, so I could not use any of these countries for my summer destination. I decided instead to put Italy on the map, a country that I had always wanted to see for its associations with Trajan's Dacia and its Byzantine mosaics.

Sicily was the place: a few years previously a strong earthquake had shook the island and several villages were wiped out. There was a local vested interest in seismic engineering, the branch of science which aimed at constructing 'earthquake-proof' structures: roads, bridges, dams, houses, in order to save human lives. The beginnings of earthquake engineering go back to the traditions of populations who always lived in seismically active regions, such as Japan, where the use of certain building materials proved to be better at resisting the effects of earthquakes. However, it was the densely populated regions of California, with the need for speedy highways with spaghetti junctions and the use of modern materials in high rise blocks, which required a new approach to building technology.

Understandably, the Sicilians were anxious to generate interest in their region and decided to create an International Summer School in Seismic Engineering, at Erice in Western Sicily.

After the fog of Newcastle, the misty Low Countries and the pinewoods of the Black Forest, this was the first truly Mediterranean spot which I had ever visited. To me this was a corner of Eden coloured by luxuriant exotic plants, palm trees and citrus trees bearing fruit, which I never tasted in Romania, as these commodities were not regarded as being 'essential' foodstuff and therefore never imported, for lack of hard currency. Even the usual fruit appeared to be better coloured in Sicily: more scented, ripe with the taste of sunshine and much larger than their counterparts sold by English grocers. My palate, still redolent with the strictures of the Communist diet, was salivating abundantly at the sight of these fruits.

EARTHQUAKES OR CARTHAGINIANS?

Erice was an ancient settlement, clinging to a steep escarpment with breathtaking views of the blue seas. The streets were narrow, like a mixture of European medieval towns with the overprint of a North African souk. There was an eerie atmosphere of the Arab world pervading, with furtive shadows of women passing in the streets, almost apologetic for coming out into the open, glaring world.

This was a man's world, the ubiquitous male, lounging in the cafés, chatting or playing games together, male youths horseplaying and bragging, but never girls! They would usually appear accompanied by elderly matrons, who would walk purposefully to their destination, looking severely at the pavement, as if anxious not to stumble over the cobble stones.

Our Sicilian hosts were proud of their mixed heritage of invading ancestors: Greeks, Carthaginians, Romans, Normans and more. Every sailor must have had his input of genes into the Sicilian melting pot, almost a paraphrase of the Latin:

'I arrived, I saw and I procreated.'

The memory of these passages was often left in stone, for future generations to acknowledge their distant past. The temples of Segesta, Agrigento and Selinunte, some two thousand five hundred years on, still arrest the visitor's imagination, with their gigantic, yet elegant Doric columns. Some two millennia on, size had an almost vulgar dimension in the Communist demonology—*bolshoy*, like the University of Moscow. By contrast, the ancient Greeks managed to give their huge temple columns a dignity and beauty, even in their ruined state.

When our international group of seismologists was taken to the archaeological site of Agrigento, the Regional Director of Historical Monuments confronted us with a key question:

"Could the experts say, simply by looking at the scattered remnants, whether the columns were scattered by an earthquake, rather than destroyed by the invading Carthaginians, during a Punic War?"

"Hard nut to crack!"

The Californian engineers had as much difficulty in coming to grips with what a 'doric' column looked like, in the first place, as to what technology the Carthaginians might have had at their disposal, to cause such havoc.

"Maybe bulldozers?"

Seeing that he did not get very far with an answer, the Director then showed us a collection of historical photographs of the same site.

"Before Mussolini moved the columns to their present position, all lined up on the ground".

Well, well! *Now* we knew what we had in front of us was no longer irrefutable evidence, simply evidence that was tampered with by the *Duce*.

Presently the experts were more confused then ever. How could they tell? As nobody ventured any solution to this riddle, I stepped forward and said:

"First the Greeks lined up the marble columns on the ground, ready for the erection of the temple, then the work had to stop, because, in the meantime, the Carthaginians had invaded and so the temple was never erected."

"See? You need not be an archaeologist or a seismologist, for that matter, to make sense of these shambles: just bring in the Romanians, as they will always come up with the right explanation!"

OXFORD, NORWICH AND NEWCASTLE

As my bank balance was in a healthier state, courtesy of my translation work, I realized that it was time to give a filip to my PhD dissertation. I was not short of ideas, but somehow I felt that I worked in a vacuum and that I needed some feedback from an independent source. Some self-doubt settled in and could not be brushed aside.

What if my ideas were wrong and I got myself in some cul de sac?

The best way to see if this was true or not, other than ask my supervisor, was to start a series of seminars at various universities and rehearse my new ideas, see how they would feel about them.

Would they tear me to pieces? Would I make a fool of myself?

Well, what better way to find out other than by going into the 'lion's den' and laying myself bare? Perform some geophysical striptease, expose my most intimate thinking on the question of *non-rigid'* plates? Until then all plates had had to be *rigid*—I myself thought otherwise and introduced two new categories of tectonic lithospheric plates—one called a *'sub-plate'* and another called a *'buffer plate'*. This iconoclastic approach, within a community which by definition was conservative, would cause at least some raised eyebrows, if not outright hostility.

"Let's go for it! What an excitement!"

I wrote to various Professors, Heads of Geology and Geophysics Departments, stating the topic of my PhD thesis, which I offered to present in a seminar. It was most unusual for a research student such as myself to go on the trot and give such talks, as this was mostly the preserve of teaching staff of some repute, Fellows of the Royal Society or managers from industry. I was none of these, but it was worth trying. In this respect my initiative was a departure from the norm.

I did not target the universities at random, on the contrary, I chose those whose line of research was similar, or whose feedback was most likely to be useful.

I wrote first to Fred Vine, formerly at Cambridge, who had just returned from Princeton to be appointed Professor, Head of the Department of Earth Sciences, University of East Anglia.

As one of the founders of plate tectonics, how would he view the conceptual changes which I was trying to introduce?

I took the train to Norwich with some apprehension. I had met Vine before and found him affable and unstuffy, which put me at ease. Norwich was not exactly on the map of the *Grand Tour* of speakers and Vine's department was new. I was gratified to hear his comments, which were not at all critical, just constructive. He acquiesced to the most important point—the non-rigidity of some plates, which implicitly acknowledged the existence of a 'different' type of lithospheric plate. To hear this from the *'Guru's'* mouth gave me some initial confidence.

To unravel what was happening in Central Asia behind the Himalayan arc, an area known to have the widest belt of seismic activity, I devised a simple technique of locating epicentres and plotting them in a different way. I used as input the body waves (USCGS) magnitude of thousands of events. When plotting these in categories of half-magnitude increments, a pattern started to appear, which showed a strong correlation between the larger magnitude earthquakes and the rhegmatic fracture pattern of the area (Tibet, Tarim, Sinkiang and the Baikal). Clearly only earthquakes above a certain magnitude threshold were correlating well to well defined tectonic features, whilst the smaller magnitude earthquakes had their epicentres distributed in a more diffuse way, over the whole area. These hyperseismic chunks of continental crust behind the Himalayas were caught up between two rigid entities: the Indian Plate to the south and the Eurasian plate proper to the north, and they were acting as a 'buffer zone' between the two. This is how the term for a non-rigid plate, a *'buffer plate'*, was defined for the first time.

I wrote to Dr Ron Oxburgh, in the Department of Geology at Oxford, thinking that a visitor from 'the other place' might not be met with the most lenient attitude, especially if he preached 'heresy'. Here the discussions were lively and the ideas caused some excitement, at the end of which I felt that I had got a pat on the shoulder and a wink. Ron Oxburgh later became Master of Queen's College Cambridge, Head of the Geology Department at Cambridge, Advisor to the Ministry of Defence and more recently Rector of Imperial College London and elevated to a peerage as Lord Oxburgh.

No sooner was my talk in Oxford given than I arranged to go north to Newcastle. It was here that I first heard McKenzie giving his fateful paper on plate tectonics and I thought it appropriate to give one myself, some two years on. It was also an excuse for seeing old friends again. Professor Creer arranged my visit and the auditorium was full. To me this was another rehearsal of ideas, an excuse to polish and further refine certain aspects of my work by presenting it in public.

Not all talks were identical, as I always spoke without notes and often changed the slides. Criss-crossing the country from East Anglia to Northumberland was like an electioneering journey, during which I had to reveal a 'party manifesto'. My 'manifesto' was one of an innovative model, meant to modify an existing concept which I found inadequate. The whole idea was to proselytize, to refine, to compare notes, to make contacts and friends, to put my name on the map of British Academia and essentially to show the courage of my scientific convictions.

"Anybody voting against 'buffer plates'?"

"????"

"No? Nobody?"

LIÉGE

None of my English seminars were paid for, only the expenses. In fact it never entered my mind that I could possibly receive a fee. I had no reason to carry on my ego trip by continuing my series of lectures, as the original objectives were satisfied: I got the confidence and the blessing, I put my name on the map and I polished my models. What else could I wish for?

On this particular occasion, however, the reason was sentimental. My girlfriend Roxana had found herself a job as a PA to the General Manager of the International Iron and Steel Institute in Brussels. This was not too far from Cambridge, but far enough for an impecunious student to remain out of physical touch.

How could I bridge the geographical gap? It was impossible and beyond my reach to pay for a trip to Brussels. Somebody else had to pay for it, but who? What if I gave one more seminar, the last one? That would pay for my journey to Belgium and back.

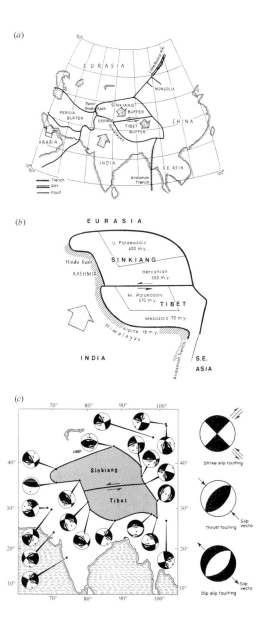

*(a) The rigid crustal plates which India and Eurasia present have trapped between them the two buffer plates of Sianking and Tibet. (b) Sketch map of the Sinkiang and Tibetan buffer plates, showing their major structural units. Notice the decreasing age of mountain-building movements from north to south, suggestive of continental accretion. (c) Focal mechanism of the earthquakes around the two buffer plates. Each mechanism is the image of the Earth in stereographic projection showing zones of compression (black quadrants) and dilation (white quadrants). (After C Roman 1973 Buffer-plates—when continents collide, New Scientist, **57** (830) 180–181.)* *(See page 95.)*

This was made possible, as I had a university contact in Liége. Jean-Clair Duchesne was the son of Professor Jacques Duchesne, Chancellor of the University of Liége, whom I first met in Romania. Professor Duchesne was generous enough to correct my translations of Romanian verse into French. By coincidence, his son was a geologist and spent a year at Cambridge, where we met and made friends. Now Jean-Clair was back as a lecturer in Liége and ready to arrange my seminar. It was the first time that I gave a talk in French, whose title was *'Sur la limite des plaques lithosphériques dans la croûte continentale'*. The Department of Geology and Mineralogy of Liége was at that time still in the old university buildings, before the new campus was built out of town. This was a journey amongst friends.

Professor Jacques Duchesne and his wife received me for supper. He was a distinguished Orientalist and an expert on Paul Valéry and also had written essays on the Romanian artist Constantin Brancusi. The Professor's wife was extremely erudite, an expert on old musical instruments, with several publications on the subject. They lived in a house with beautiful gardens, on the top of the hill overlooking Liége. We talked about Romania. Their geologist son, Jean-Clair, had recently married Claire Goffin, a paediatrician.

The Goffins were keen art collectors and they lived in a wonderful Bauhaus villa outside Liége, full of modern sculptures and paintings of the inter-war period. Claire and Jean-Clair had built for themselves an architect designed villa, in Tilff, also on the outskirts of Liége, in a style reminiscent of Le Corbusier, with very high ceilings, wide spaces, glass partitions and terraces, which made the surrounding woodlands virtually merge with the living room.

In the evening, Jean-Clair arranged for two of his colleagues and the Professor of the Department to join us at a traditional dinner in town, of 'moules et frites'. This was a very convivial meeting prior to my giving the lecture. In the lecture theatre the 'feel' was rather more Continental, with a touch of Anglo-Saxon informality. In the first row was seated an octogenarian gentleman who might have descended from one of Ramsey's canvasses—an exiled Austrian aristocrat, Count Radetzky, whose ancestor, a General in the Imperial army, was better known from the *Radetzky March*, dedicated to him by Johann Strauss the Elder. Count Radetzky made the point of attending all geology lectures at Liége University, whether he understood them or not. He was the epitome of the 18th century nobleman, who might have pioneered entomology, or even geology, for that matter. He might have been a contemporary of Buffon, or Cavendish. No sooner did the lights go off, to show the series of slides, than Count Radetzky fell asleep. He did not snore, but his head lay gently to one side, quite touchingly. When the lights were switched on again, he slowly came back to life. It must have been some incredible therapy for the old boy to chase away his loneliness and still appear to be in the centre of 'high thinking'.

At the time of my talk in Liége, plate tectonics was still looked upon with some sort of reverence—not many people on the Continent were involved in such research, other than Le Pichon in France, so the topic was quite novel and Jean-Clair was very proud to sponsor it. He gave a short pep talk of introduction, presenting my academic credentials to the audience.

Very much to my surprise, the following day the Department not only paid for my expenses, but also thought of giving a nominal fee, which for me, as a student, represented an important sum. I went straight into town and bought myself a new pair of shoes. Like Cinderella's shoes, which took her to the ball, my Liége shoes took me to Brussels, where I made a surprise visit to Roxana at the Iron and Steel Institute. I never thought plate tectonics could also fill in sentimental gaps. It did on this occasion and I was immensely pleased with my enterprise. The plate tectonics conference in Liége was to augur well, because it also resulted in my proposing to Roxana in Brussels. A few years later, Professor Jean-Clair Duchesne became Head of the Mineralogy Department at Liége University and attended, as godfather, the christening of my eldest son, Corvin, at Westminster Abbey.

IMPERIAL COLLEGE AND CAMBRIDGE

Following this early success at Norwich, Newcastle, Oxford and Liége, I set my eyes on Imperial College in London. This had a worldwide reputation and it was good to add its name to my list of seminars. Again, as on previous occasions, my ideas were not challenged and I grew firmer in the conviction that I had a valid point to make, which would be credible to discuss in my PhD dissertation.

Now I was ready to give my views closer to the *'lion's den'*, in Cambridge, where McKenzie was the unchallenged expert in the field. His reputation grew by the day, as he was active with his trans-Atlantic academic contacts, which were a constant source of inspiration to him during sabbaticals and summer vacations.

I first approached W B Harland, in the Geology Department, as I wanted to give a talk on Tibet's seismicity. Harland was a reader in tectonic geology and an expert on Tibet. At the time of my studies in Cambridge there were four separate departments of Geology, Geophysics, Mineralogy and Geography. There was very little contact between these departments and to cross the threshold between them required some self-assurance: it was simply 'not done'! Furthermore, as geophysics had its own maverick, in the person of McKenzie, Geology too had a counterpart, in the very person of John Dewey, later to become Professor of Geology at Oxford. There was a friendly rivalry between the two talents, each of them very jealous of their patch. They both had teaching qualities, but Dewey's style was more inspiring to students than McKenzie's, who was more candidly mathematical and duller. In fact, during the late 1960s and early 1970s at Cambridge John Dewey's work put much needed geological flesh on the more abstract geophysical concept of plate tectonics. John's seminal thinking covered wide areas from the plate tectonics of geosynclines to Appalachian mountain building and the conversion of continental margins from Atlantic type to Andean type.

I felt a moment of trepidation at being able to address geologists in the very department created by Adam Sedgwick, a pioneer of world geology, who laid the foundations of modern geology, identifying crucial formations such as the Devonian and Cambrian. The seminar was given in a room within the very precincts of the Sedgwick Museum.

I hardly ever looked at the audience as I gave my speech. I focused my attention at the sequence of slides and if I turned to the audience, I usually looked at some distant point, with the public image blurring in an indistinct haze—I did not notice anybody in particular. On this occasion, however, for a split second, I caught sight of Harland, who had his eyes firmly closed.

Was he asleep or was it a way to concentrate on the talk, like music enthusiasts at concerts?

This was not a concert and I thought that a slide could speak infinitely more than the comments I was making. I found his perceived indolence slightly disturbing, if not outright irritating, but I carried on regardless. At the end of my talk I had surprisingly few comments compared with the audience at Oxford. Harland, however, made up for the rest by asking some very pertinent questions, as if he had never had a nap.

It was Lent term 1972, and I was coming close to putting the final touches to my dissertation. By now I felt that 'the moment of truth' had come when I should talk about my research in the very department where I belonged. Quite unusually and gallantly Teddy gave me a 'flower', by including me on the prestigious list of *Geophysics Colloquia*. Here I was the only student speaker, the other guests being W B Kaula from Caltech, Dr D E Cartwright from Scripps, Proffsor J E Nafe from Lamont and other high calibre world scientists. Furthermore, Teddy was to make sure that his presence at his supervisee's talk meant that everybody else was going to be present. The event had a certain *piquanterie* about it, as only two years previously I had been urged to give up the subject of plate tectonics and not take Bullard as supervisor! By this time the series of previous seminars given throughout England gave me enough confidence to give a credible picture of what I considered would become '*the New Testament of plate tectonics*'. Carol Williams, a colleague from the Geophysics Department, asked if she could introduce my newly defined buffer plates of Sinkiang and Tibet in the new BBC television programme, *The Restless Earth*.

Now I was 'made', this was the crowning of my efforts and the green light to the final touches of my thesis. From now on, I thought, this was going to be plain sailing. However, I was only lulled into a false sense of security, as I was soon going to find out, with a thud.

WISE MOLLY

It was natural that, quite apart from my research, I should get on with my other activities at Cambridge, into which I plunged myself with great zest. I played squash three times a week, I went on long cycle forays outside town, I attended parties, I went to conferences, concerts, met very interesting people in and out of University. I was also busy making a little money with translations and being a guide in Cambridge. Also, my administrative situation, stemming from my Romanian passport, required constant attention and was rather time-consuming too, almost a full-time job.

With all this happening, I was not very often in the Department of Geophysics and I worked mostly in my rooms at Peterhouse. This choice put me somewhat at a disadvantage, as I could not listen to the ramblings which were going on. Apart from a few geophysicists, like Bill Limond or Anton Ziolkowsky, with whom I played squash from time to time, I did not get many snippets of gossip from Madingley Rise. On the whole my main sounding board and guardian angel was Molly Wisdom, the ex-wife of a professor of philosophy at Trinity and secretary to Sir Edward Bullard. Molly knew all the ins and outs, Molly knew everybody, heard everything, judged (usually quickly and very correctly) everybody and she was fierce if anybody crossed her path. *But* she had a heart of gold and she took me under her wing.

I often went to see Molly in her little house, off Parker's Piece. She used to give champagne breakfasts for the whole department at her house, which turned out more like brunches, lasting well into the afternoon. Molly advised me what tactical errors I should avoid, thus providing me with a much-needed lesson in politics, never my speciality. Molly did foresee the thunderstorm approaching, but for me the sky was clear and I felt I had nothing to fear.

Although I had given lectures, I hadn't published much since my *'Nature'* article, other than an article in the Proceedings of the Luxembourg meeting and in the (quite obscure) *Revue Roumaine de Géophysique*. Publishing would have required a lot of time and I had very little of it—I had to press full steam ahead with my dissertation, seemingly a never-ending task.

"I must get you to appear in the *Cambridge Evening News*", Molly said.

"What on earth for?"

"It will be good, as a PR exercise, for people to see, for people who do not understand your work in college and so forth."

"Why not?"

I rang the feature editor of the *Cambridge Evening News* and within days my picture, in my rooms at Peterhouse, in front of a tectonic map of Eurasia, was blasted on half a page of the newspaper. It was quite a laugh, but it did the rounds of Peterhouse.

In fact, more seriously, this was an early alarm signal, which I did not understand at the time, but the danger was not long in materializing and took me by surprise.

Photograph taken in my college rooms at Peterhouse for the feature article which appeared in the Cambridge Evening News in 1973. In the background is the tectonic map of Eurasia, on which the Tibetan and Sianking buffer plates are clearly visible.

MIT SALVO

Molly's intuition was right. Maybe it was not 'intuition' on her part, just 'inside knowledge', which she could not divulge, and so she tried to 'save' my work with an article in the *Cambridge Evening News*(!).

The latter would hardly suffice to save my research skin: there were rumblings and I was soon going to be in deep trouble: Teddy called me at Peterhouse, saying he wanted to see me. He had never done this before, and it was serious. He had a long face.

"You know, Constantin, something bad has happened. I told you to hurry up and finish your thesis. Now there are these fellows in the States, who are doing exactly the same research as you are doing here, on the Central Asian earthquakes, and they have sent this manuscript for publication. The article has been approved and it will be in print shortly. This means that all your work will lose its original aspect, as it will have been published before, by somebody else."

In this particular case the author was Peter Molnar, whose fame had travelled fast in the world of plate tectonics. I looked blank, as if somebody had hit me on the head with a sledgehammer. I felt numb and completely speechless. I looked at the publisher's proofs of the article, which Teddy had in his hands—exactly the same region, the same earthquakes, the same focal mechanisms and exactly the same fault plane solutions. There was no way in which I could write something myself and have it published in a scientific journal, without waiting for six to twelve months. Molnar must have submitted his paper at least six months previously.

Suddenly I felt very angry: it was the supervisor's responsibility to see that nobody else at another university 'trespassed' on a student's area of research, in order to secure the integrity of his work. But Teddy did not do it, he could not do it, because he was less in touch with the subject and with what other universities were doing in plate tectonics. This code of practice had been broken, but I could not be angry with Teddy—he had acted in good faith and he did not know.

Eventually, when Dan had broken the news to Teddy, it was too late for me to take protective measures.

I asked Teddy how he had known about this article.

"You see, Dan was asked to be a referee."

"Ah, I see."

Teddy guessed my thoughts.

"You know", he said, "Your work on the Carpathians could secure you the equivalent of a Master's degree. I could write to the Registrar and explain the situation and propose you for a Master's".

The old man meant well, but for a minute I thought he was out of his mind: how could he suggest that I should do a Master's, when I came here for a PhD? After all, it was an established practice for somebody who failed his PhD to be given a Master's at Cambridge. This was an insult. I clenched my teeth and told Teddy that I would think about it.

I was a slow thinker, I needed time and I was exhausted.

THE BRAIN WAVE

It took me ages to reach Peterhouse from Madingley Rise. My knees were wobbly and I had to push the bicycle: the sky had fallen on me and I felt as if I was in some dark hole. I went into my rooms in Peterhouse and I paced up and down, like an animal trapped in some zoo enclosure. I had to do something, I could not believe that such a thing could happen, that it had happened to me. It was altogether preposterous, wasteful and spiteful. I must not let the other side have the satisfaction.

But *how*? Quite simply by publishing an article before Molnar's. But *how*? This would be an impossibility, because of the inherent time handicap: I couldn't publish it in a 'quarterly', not even in a 'monthly', even for *Nature* it required several months.

Were there any journals which might accept something quicker?

My mind scanned feverishly, but in vain. I went to bed exhausted and without hope.

In the morning I went for a hearty English breakfast of bacon and eggs in College—I hardly ever took a proper breakfast, as I used to have instead a large mug of Turkish coffee, which I would prepare in the hostel. I felt I needed some sustenance in order to think straight.

The Hall at Peterhouse was always lovely, with the high church-like ceiling, with massive carved oak beams in mellow shades of brown, with wood panelling, the stained glass and stencils by William Morris, and the old portraits of Masters and 19th century luminaries, including a painting of Cavendish by Zoffany. It was a joy to restore my spirits in this environment and I felt better for it. My mind, though, was still in search of a solution to my intractable problem and I soon found myself pacing up and down my rooms in College, whilst drinking my statutory mug of strong, black Turkish coffee, which was like dynamite.

I scanned again the sequence of events from the first day I arrived in the Geophysics Department at Cambridge. This was almost three years ago: my first readings, my computations of Carpathian earthquakes, the *Nature* article (what a coup, I thought it safeguarded at least a portion of my research) and the review in *New Scientist*.

What? *New Scientist*? Why didn't I think of it before? Silly me! How very silly of me! Gosh! I must go straight to them. They published weekly and I would have no problem getting into print before Molnar *et al*'s article was published in the States.

There were two difficulties that I could foresee: One was to convince the Editor of *New Scientist* that the topic was of interest. The second difficulty was to write the article in a format and at a level required for the readership, without as such compromising the essence of my research. Nothing else mattered, provided that they agreed to include *one* crucial diagram which summarized thousands of readings of microfiches carried out at Frankfurt and processed at Aldermaston and Edinburgh in the preceding years. This represented the core of three years of work. It was my diagram with focal mechanisms of Tibetan and Chinese earthquakes of magnitude larger than 5.0 on the Richter scale. This was a scale of earthquake strengths, marked from 0 (the weakest tremors) to 10 for the largest ones. Destructive earthquakes on the Richter scale start at magnitude 5.5. As the sensitivity of seismographs has much improved since 1935, when the scale was devised, there are now measurements of smallest magnitudes, which are marked 'negative'. Charles Francis Richter (1900–1985) worked at the Carnegie Institute and California Institute of Technology. His co-worker was Beno Gutenberg, who calibrated the scale of earthquake magnitudes on Californian events. Gutenberg and Richter's scale (which is the correct name) is a logarithm of the maximum amplitude of earthquake waves, observed on a seismograph, adjusted for the distance of epicentre of earthquake.

The diagram which I tried to safeguard through publication in *New Scientist* showed, for the first time, the outline of Central Asian 'non-rigid plates', which I had defined and called *'buffer plates'*.

I went to the public phone box in College and rang the London office of *New Scientist*. Yes, Peter Stubbs was there.

"I am Constantin Roman from Cambridge. Do you remember me? You reviewed my article in *Nature* on the Carpathian earthquakes."

"Yes, I do remember quite well. What can I do for you?"

"I am in a spot of bother. I am doing this piece of research on Central Asian earthquakes—you know? It is a bit of a rat race."

"I know."

"Well, I just discovered that there is this American, from MIT, who has done the same research as me, quite independently, and came to the same results. He is about to publish his results in a scientific journal and, of course, if he publishes before me, my work is ruined. I won't be able to present my dissertation as an original piece of work. My work of the last three years will have been in vain."

"I understand."

"Good. Now I think that the subject is so 'hot' that it will certainly interest your readership. Besides it will be a world 'first'. You will have published this piece of news before anybody else."

"Certainly, I can see that."

"Would you be interested?"

"Yes, we are, provided that a non-specialist readership will understand and the text will have to be quite explicit, void of jargon terms or, at least, these will have to be explained."

"It goes without saying, I understand. How many pages will you give me?"

"It will have to be a feature article. Are there any illustrations?"

"Yes, there are, one or two, but one in particular is crucial in securing the originality of my work. How many words do you want me to write?"

"Six thousand words."

"You do not mind if you will have to correct my English?"

"No, this is no impediment."

"How soon can you have it published?"

"We always work three issues in advance. If you come with your manuscript this Thursday we can get it printed in two and a half weeks."

"Good, I am overjoyed. I will see you this Thursday in London. A last question though, the diagrams will have to be drafted, have you got an illustrator at hand?"

"Yes, we do."

"I will have to explain the drawings to him when I come."

"He will be there."

"Good, thanks. Thank you very much—you have saved me from a terrible quandary."

THE *NEW SCIENTIST* LIFE LINE

It was Tuesday lunchtime and I had only 36 hours before I would have to hand over the article to *New Scientist* in London. I was so excited about my brain wave that I was rubbing my hands and any undergraduate looking at me must have thought that I was potty. Never mind that—Cambridge was full of eccentrics and oddballs. One more would not make any difference. The essential thing was that I could win the race and now this possibility became real. My state of stress was such that I could hardly cope with the unfolding of events—now you succeed, now you don't, now you succeed again.

The rest of Tuesday was completely wasted in frantic speculation about what happened, why it happened, how was it possible, what might happen next?

I could not wait: now the initiative was mine again and whilst the insidious opposition might have relished the thought that I was licking my wounds, I was well away, lining up my cannons.

The following day, Wednesday, I frantically wrote my 6000 words, typing with two fingers on the inadequate, unwilling, cheap French typewriter which I had bought in Paris. My wastepaper basket was full with rejected drafts, as it was very hard to compress the essentials of three years research into a short article and explain it in the simplest possible manner. It was an excellent discipline and eventually I managed to distil enough to put together a good story, without compromising the essential details. The following day, I took the early train to London.

It was the first week in January 1973 and London was wet in the drizzling rain, with a grey overcast sky. People were hurrying under their black umbrellas and I went straight to the *New Scientist* offices, clutching the manuscript under my arm.

Peter Stubbs received me immediately and we went about our business. I explained first the principles of the subject and the importance of its novel ideas to the advancement of plate tectonics. He read the manuscript and gave the verdict.

"It's good, we will put it in the next issue, now let us see the diagrams."

He called the illustrator into his office and I explained what was important to retain in the figure with focal mechanisms. We agreed to retain only three sketches in the text, two of which showed the boundaries of the newly defined buffer plates—Tibet and Sinkiang—being squeezed between the Himalayas and Siberia, with India pushing in a north-easterly direction. I went back to Cambridge and carried on working in my rooms in Peterhouse, avoiding the Geophysics Department until the article was published.

On 25 January 1973, I went to the nearby newsagent in Trumpington Road and scanned *New Scientist*, to check out how it all came out in print.

"There it is, there it is!" I shouted—the newsagent raised his eyebrows I was usually a quiet man.

"What on Earth hit him?"

"Here it is!", I said, showing it to him—he was the first person I could show it to. He shrugged his shoulders. I bought all the other copies of *New Scientist*—now he smiled—he had a good turnover. I could not wait. I could still get to the department in time for coffee at 10.30 am. It was 10 am. I jumped on my bicycle and covered the distance to Madingley Rise in record time.

Teddy's office was on the ground floor, to the left of the main entrance, off the hallway. His room was large, overlooking the lawns and garden shrubbery, which opened to the paddock beyond, where horses were grazing. It was a peaceful, idyllic scene. Teddy's office was bursting with publications, on endless ranges of bookshelves. Manuscripts of current work covered his desk. He had his large desk at the far end of the room, so he could see through the window and also see the visitors who were ushered in by Molly Wisdom. As I was 'one of the family' I had no need to be announced, I went straight in, waving the issue of *New Scientist*:

"I have some wonderful news for you, Teddy."

I stopped. I was surprised at myself—it was the first time in three years that I had called him by his nickname. He must have been surprised too, because he smiled and asked:

"What is it? Show me."

I pointed to the article in *New Scientist* with the diagrams of focal mechanisms.

"I have made it, I've beaten the clock", I said. "I am saved. I published before Peter Molnar! Now everything will be all right won't it? I can take my PhD!"

Teddy agreed—my enthusiasm was contagious and he was happy to see his student succeed. He was impressed about my resourcefulness at winning the race and securing the originality of my ideas. From now on everything was going to be plain sailing.

We went together into the Common Room to have coffee and I distributed all my copies of the journal for the other geophysicists to see. They knew about my difficulties and about the misfortune which had befallen me and they had watched developments with interest, almost like a betting game—to see who wins. Did they ever place bets? I very much doubt it, but if they did, the odds would have been against me. Now the balance had tipped the other way, and I was a favourite again. They could welcome me in their midst without embarrassment. Nobody likes losers.

Drum Matthews was there too and he joined in the discussion, laughing at this latest twist of events.

Dan McKenzie usually came a bit later, when everybody was already in. Teddy took Dan aside and explained the outcome of my publication. I presume he did so not because he was under any obligation, but simply because Dan was the referee of Molnar's article, which was yet to appear in print.

Molly was pleased too. We had won the second round, in three years—not bad, not bad at all, for a Cambridge match!

TRANS-ATLANTIC REPARTEE

I rang Peter Stubbs, the Deputy Editor of *New Scientist*, to thank and tell him how Sir Edward Bullard had received the article. I asked Peter to send me one hundred reprints of the article, to post to other plate tectonics specialists at universities on the Continent and in the States. I wanted in particular to send a copy to Peter Molnar.

"Don't worry", he said, "It's done, I have already ordered them."

"How kind of you."

"It's a pleasure."

Now I could despatch my news to Peter Molnar at MIT.

By this time, Teddy was at the University of California in San Diego and he wrote to me, on 14 May 1973:

"I was also glad to get your paper in *New Scientist* which seemed to me to read very well. Peter Molnar asked if you had seen his MS before writing it, I said you had, but assured him that you had the focal diagrams long before that. He was surprised how well the focal mechanisms agreed with his, but I think I made it quite clear that yours were obtained independently. I have written to Reading. No more now. I have to go and give a lecture, Yours, Teddy."

This sums up how delightfully generous Teddy was. He was thousands of miles away and very busy too. Yet at the same time he was thinking of his student, back in England, reassuring him about the content of his thesis, commenting about his latest publication, sending a reference for a job application and at the same time keeping him abreast about the reaction of his research competitors!

Were there many academics to match his stature, I wondered.

Another instance of Teddy's tact in leadership proved to be his reaction on finding out the omission of my work in the *Reporter*. Each University Department at Cambridge produced an annual report of activity, which was published in the *Reporter*. This was an abstract of the main advances made in various areas of research, publications, seminars given by various members of staff and students, visitors or perhaps honours degrees received. I never paid much attention to these, although I had the right to at least one line in the first year.

By my third year it dawned on me that it might be useful to have my work fully acknowledged in the *Reporter*. At this point in time I had to my name four scientific papers and a good numbers of seminars. I sent the full list, thinking that it would be good for my Tutor in Peterhouse to see that I was pulling my weight in difficult circumstances and that his efforts in helping with various administrative tasks, on my behalf, were fully justified. By my third year, correspondence with the Romanian Embassy and the Home Office, over my status in the UK, was reaching alarming proportions.

Very much to my surprise, I noticed that *all* my activities in the Geophysics Report were completely left out. I simply could not believe this could happen to me, at the crucial moment when I needed most help. Yet everybody else had their smallest contributions and trips put in. Molly was on holiday and I could not ask her. Teddy was at La Jolla and out of reach. Somebody else was holding the helm in Teddy's absence. As I was in regular touch with Teddy, in La Jolla, I asked for an explanation. I could hardly reproach him that, in his absence, somebody else had completely 'forgotten' about my input and left it out altogether. I simply said I was disappointed that my work was ignored in the *Reporter*. Teddy wrote back a very nice letter apologizing, as he felt that he bore responsibility for this omission, although he was absent at the time: he said that my work was 'omitted, not excluded!' Teddy added, tactfully, that in the previous year's *Reporter* another of his students was, inadvertently, omitted.

How generous, graceful and very tactful Teddy was: I felt enriched by him in more than one way, not just by learning geophysics, but also the civility of his *savoir faire*.

IVAN THE TERRIBLE—ROMANIAN STYLE

Life at Cambridge became rather hectic. I felt sometimes that I had taken on too much. But I had no time for a respite or self-pity. Yet things were not easy, were not easy at all. My parents in Romania were not allowed to visit me, as the Romanian authorities felt that I had stayed too long in England, and they were annoyed by my Tutor in Peterhouse, who was constantly writing on my behalf to the Embassy.

The British authorities were not any easier either, as my time in Cambridge was soon going to be up and I needed to clarify my resident's status in England. The correspondence with the Home Office reached alarming proportions and the whole logistics involved in getting support was phenomenal.

To make things worse, prospective in-laws, who discouraged any contact with a penniless Romanian, further complicated my emotional life. This led to the girl I was courting not being sure herself.

Money was short, on a £600 per annum scholarship, intended for three terms, I had to live on it for 12 months. At £50 a month I was getting by, but I had to supplement my income through extra paid work, such as translations and being a guide in Cambridge for a travel firm, but all this taxed my valuable time.

I still had to complete the text of my dissertation and find money for producing it (secretarial, photocopying, graphics, etc). It came, therefore, as no surprise to Teddy that although I was near completion I needed a fourth year.

But where to find the funds? Peterhouse awarded only a three-year scholarship and though quite a few research students needed a fourth year, this was not encouraged. A lot of postgraduates who didn't finish in time carried on with personal funds. A few became 'professional students', as on the Continent, and those few who tried usually failed their PhD and ended up having, at best, an MA.

Teddy approached Peterhouse, who took up the matter at a Governing Body meeting in College. It was a difficult one, as in addition to providing the funds, there was a considerable effort by the College in making representations on my behalf to the British and Romanian authorities. My file was probably the largest in the College archives.

"Better get shot of him. But how could one?"

There were those amongst the College Fellows who did not see why the goodwill should be stretched to a fourth year, for purely financial reasons. Yet the College was a registered charity and was known to be well off. This minority of Fellows was also the most vociferous:

"Could he not become a teacher?"

My Tutor, Roger Lovatt, passed on the suggestion to me.

"Couldn't you become a teacher, instead of finishing your PhD?"

"No", I said, "I have come here to do a PhD and I will see that I will do it, come what may. By the way, isn't the College a charity?"

"Yes, it is."

"But some people are *more equal* than others?"

Bullard had witnessed this situation before. He was used to it in Cambridge, and was not impressed. He told the College Governing Body that he would offer half of my Scholarship funds for a fourth year if they found another £300 to match it. My Tutor came to realize that I was determined and that quite a few ripples might be caused if the funds were refused. I remember my quoting to him, as a historian, the saying of a 16th century Moldavian Prince, who told his unwilling Parliament of Boyars:

"Dacã voi nu mã vreţi, Io vã vreau." (If you do not want me, I want you).

This was the Romanian version of Ivan the Terrible (*Ioan Vodã cel Cumplit*) and I felt as such.

Bullard's request was granted, so I was safe and sound once again, but pressure was mounting to finish the dissertation.

Molly, whom I would consult often, over an evening glass of sherry in her basement sitting room, would press me too:

"Finish it, finish it. It is important. Once you get married you will not be able to."

Marriage was a long way off, with no degree, no money, no permission to stay in England and no job, how could I get married? It was crazy. Besides, my future mother-in-law decreed that I could marry her daughter "Only when you become a British citizen."

The old bird knew that I would require at least five year's of residence before I could be naturalized and she hoped that, by that time, my enthusiasm might wither away.

So much for that!

In the meantime I had to take one thing at a time and for the moment the dissertation had to be top priority.

LETTER TO THE *JOURNAL*

I had completed the first part of my thesis on the Carpathian earthquakes, had most of the Central Asian chapters done and submitted the draft manuscript to Teddy, who made useful suggestions, then he added the usual sentence:

"But you must finish."

"Certainly I will."

I was still looking for the latest publications in my subject, in order to bring everything as up to date as possible.

During my search for the latest scientific articles, I came across a reference to the *'Active tectonics of the Mediterranean region'* published in the *Geophysical Journal of the Royal Astronomical Society* in London. Although the subject dealt with the Mediterranean it also made crucial statements on the Black Sea, the Carpathians, Transylvania and the Pannonian Basin, which I felt, rightly or wrongly, was 'my preserve'. The paper made ample reference to my key contribution in *Nature*. I did not agree with what it said, but I could not ignore it, because of its overlap with the topic of my dissertation.

From the research, which I carried out over large tracts of seismicity over the continental crust of Eurasia, one aspect appeared self-evident. The breaking up of the Eurasian plate into a mosaic of smaller 'microplates' was not a credible solution. Defining arbitrary plate boundaries within the wider pattern of scattered epicentres was a contradiction in terms. Any 'microplates' which might ensue from such an exercise ran contrary to the meaning of 'global tectonics'. The practice was establishing a dangerous if self-defeating precedent. Often two different authors will come up with very different solutions of new microplates in the same area: worse still, an author might amend the shape and size of a microplate, at very short intervals, from one publication to another, for no apparent tectonic or seismic reason. I went to see Alan Cook, Jacksonian Professor of Natural Philosophy in Cambridge (and as such Dewar's successor, my 'old friend', painted by Orchardson). Alan was later to be elected Master of Selwyn College. Professor Cook was a Fellow of the Royal Society and an editor of the *Geophysical Journal*. I gave him my comments and I assume he was quite amused. He published my letter to the journal in the following issue. This was an unexpected bonus, which I received with relish.

By this time I was too busy finishing my dissertation to indulge in the secret pleasure of anybody's discomfiture over the inadequacy of the plate tectonics concept. Quite apart from the mischievous aspect of my repartee, the *'Letter'* to the *Geophysical Journal* provided me with a most welcome opportunity to publish yet another article in which I could present my latest views on the definition of 'buffer plates', 'rigid plates' and 'sub-plates'. Eventually my strong objections to the proliferation in the number of small rigid plates in the Eastern Mediterranean were vindicated, some 20 years on the Aegean plate 'disappeared', through lack of scientific evidence, and the 'Transylvanian plate' never really got off the ground. My personal disadvantage, in this measure of iconoclasm, was that I made my objections too soon. At that particular time I was a dissonant, if lonely voice: I may just as well have been put on the pyre for the strength of my unorthodox convictions and I was ready for the sacrifice.

OVERQUALIFIED
It was time to think of getting gainful employment after my graduation. This was not simple for several reasons. Firstly, in the early 1970s we were again in one of the troughs of employment, which periodically bedevilled the geological world. Geologists were practically unemployable and geophysicists, by association, were also doomed. Secondly, I had an added difficulty in requiring a work permit, which made potential employers shy of offering a place. Thirdly, I needed permission to stay in the UK, which was impossible on a student ticket, and without which I could not obtain a work permit. Funnily enough the opposite was equally true—I could not be given permission to stay in England without a work permit. Some masochistic contortionist, somewhere in the recesses of the Home Office in London, dreamed up this vicious circle and it worked wonders. Nobody like myself could get either permission to stay, or a work permit.

I had tried, over and over again. I had sent over 160 applications, not just for jobs with oil companies, but in computing, business management and universities. I even tried Marconi and related defence industries, but their doors were firmly shut to passport holders from Communist countries. I suppose they were even amazed that I had tried in the first place, but it was a pure waste of time. However, they politely answered my applications, with some nondescript, innocuous excuse. In most instances one line was recurrent: 'you are overqualified'. Maybe I was, by British standards, where most graduates finished with a BA at the age of 21 and went straight into industry. I was 32 and had no industrial experience, a very unusual candidate, whom most companies did not know how to deal with, other than reject.

All of a sudden, I felt almost unemployable. My letters of rejection piled up, one after another. Some companies did not even bother to answer. Others, as was the case of British Petroleum, told me 'not to apply (to them) anymore'.

By this time I was quite exhausted and dispirited. Teddy got rather anxious for me too. Was there a solution in sight? He was ready to write, and he did, to anybody I could suggest, just in case. Good old Teddy, he was very worried for me. He told me that when he graduated from Cambridge in the 1930s, during the Depression, Rutherford told him to 'take any job'.

I got the message. I had to take *any* job. This was an order from Rutherford himself! How could I do otherwise? But I had no joy, not for lack of trying. Only foreigners offered me a place—the Germans and the Canadians, but Roxana did not want to be an expatriate and I had to take her wishes into consideration.

It was a Catch-22 situation, like many others I had lived through before. What gave me strength was a fairytale from my Romanian childhood, where the Prince was forced to undergo tremendous trials of self-sacrifice, in order to obtain the hand of the Princess. There was no place for disillusionment.

Maybe, as nobody offered employment, I should create a job for myself. I knew how successful Oxford had been with their Laboratory of Physics applied to the History of Art and Archaeology. Cambridge too might do the same. Cambridge had the Fitzwilliam Museum, which was situated next door to Peterhouse. I could see 'the Fitzwilliam' from my rooms in the hostel, across the road, in Trumpington Street. 'The Fitzwilliam' could do with such a laboratory too and I could be the leader of it. I went to see Michael Jaffé, the director of 'the Fitzwilliam', a Professor of the History of Art and a member of the advisory council of the V & A Museum: very polite, very affable, and almost eerie. Did he understand what I was talking about? I hoped he did. I talked to Ted Hall at Oxford, I even ventured to ask Roxana if money from her trust could be put towards creating such a Laboratory—'and we would be happy ever after in Cambridge'. Unfortunately nothing came of it. It took another ten years, or longer, for the Fitzwilliam Museum to acknowledge the need for such support from science. The outcome was the Hamilton Kerr Institute for the History of Art, at Whittlesford, near Cambridge. My idea was excellent, but the initiative was premature.

I buried myself deep into hundreds of application forms and support references. Some training on how to write letters and organize a curriculum vitae might have saved me a lot of trouble, but I learned, the hard way, as I went along.

Still, I was not to know and in the meantime I carried on regardless, clocking up more and more applications.

Large organizations took weeks to acknowledge receipt of correspondence and even longer to give a final verdict. This was usually a perfunctory formula, such as 'lack of experience'. Of course, coming straight from University I would have had no prior industrial experience: my applications were for a junior position.

A doctorate was not a prerequisite in industry, which is why some employers ventured to use as a reason for turning me down the fact that I appeared to be 'overqualified'.

At some point I nearly lost track of how many jobs I applied for, as I gradually relegated the rejections to the dustbin.

HOSPITAL BLUES

Teddy was constantly writing supportive references and closely followed the progress of my applications. From time to time, he consulted and made suggestions as to what else I might try next.

Poor old Teddy, he could not impress sufficiently hard on me that I had to be 'less choosy' and 'apply for anything'.

Perhaps the most telling, if poignant example of Teddy's concern for his student's well being was his letter from hospital, which he wrote on 19 September 1972:

"Dear Constantin, As I expect you know, I am in hospital. If you would like to send me bits of thesis I will read them and send comments. I am likely to be away for 4–6 weeks. Nothing v. dreadful is wrong." Then he proceeded, in truly scientific detail, to give me a graphic description of the tubes and sundry contraptions inserted in his body and eventually ended by saying: *"I should be as good as ever in a month or so. Operation I suppose will be next week. Yours ever, Teddy."*

This touching note required an appropriately 'filial' response to my 'surrogate father' and stalwart friend, to whom I wrote affectionately:

"My dear Teddy, I was extremely sorry to learn that you were in hospital and now that I heard that you successfully underwent the operation I wish you a speedy and complete recovery. It is extremely kind of you to think of me in these circumstances, but there is nothing to worry about, so far as my dissertation is concerned and after all it would be too boring to read my prose whilst in hospital.

I have some good news for you: I approached Lord Blackett and explained my dilemma and he promised to take up my case with the Royal Society but firstly he would need a full account in writing. I will show it to Drum, before I send it off to Blackett. There is also renewed interest in my being employed by Imperial College and to this effect I talked to Professor Sutton, Mason and to Dr Ambreseys. I also contacted the Scott Polar Research Institute for a possible opening. So, you see, I have not been idle and these new prospects keep me happy enough to relaunch my more vigorously writing the dissertation.... Do get well soon! Yours ever, unmistakingly Continental, Constantin."

Nothing came of any of the above attempts, but on occasions I had to reassure Teddy that I applied for work, not just in geophysics, but in fields as remote as journalism and finance. I told him that I had even written to Jim Slater, then the whiz-kid and darling of the City, but nothing came of it. Slater was the only person who took the trouble to respond to my letter *by return* asking me to an interview.

Jim Slater's financial empire eventually collapsed, like a pack of cards, but twenty-five years on I have a greater respect for Slater, for having 'spotted the talent' and for having given me the 'sporting chance' of an interview. This was a far cry from the behaviour of the 'big boys' in industry, who were slow in responding and arrogant. More often than not, the latter group considered it to be an 'honour' even to be asked to an interview, let alone to be offered employment.

With the economic decline of the 1990s, a lot of the established companies have been taken over, cut to bits and sold piecemeal, cut to size, or made bankrupt and relegated to the dustbin, where they rightfully belonged. But at the time of my applications, in the early 1970s, I was not to know and I would have been grateful to anybody for making any offer of a job, within reason. They did not!

I told Teddy that I had also applied for jobs abroad, as I had no intention of staying on, unemployed, in Britain. I reassured him that I was

"Ready to go to the ends of the world, except to a Communist country".

A ROMANIAN DEGREE'S WORTH
One of the inherent difficulties, even for the most benevolent potential employers ready to confront the intractable question of a work permit, was making sense of my Romanian academic qualifications:

"Tell us, Mr Roman, what is your Romanian degree worth?" was a recurrent question.

If the interviewing panel was nice and I liked the position I was applying for, I would answer painstakingly about every single detail which might help them equate my degree with an English one. If I did not care and found my interviewer obnoxious, I would respond with a familiar quip of insolence:

"I could not tell: I have been in Cambridge four years now and nobody asked me this question".

Hardly the answer to endear oneself to a potential employer, the outcome being predictably negative. Sometimes I was beyond caring, yet I knew I was shooting myself in the foot, when in fact I needed all the help I could get.

FIVE POUNDS FOR AN MA
One such occasion was at the University of Leeds, where I applied for the position of Assistant Lecturer. From some sixty candidates I was on a short

list of six, which eventually was further reduced to two candidates: I was one of them and was amazed by this unusual piece of luck.

It took ages to reach Leeds from Cambridge, changing trains several times. The city was grey and the University greyer still. I was finally ushered into the Council Room, presided over by the University Vice Chancellor, who predictably asked me:

"What is your Romanian Master's degree worth?"

I did not think twice about what I was going to say, as by this time I had rehearsed the answer. I started first in a considerate tone, explaining the parallels between an English BA and MA and an equivalent degree in France or Germany, hoping that the panel might be more familiar with the qualifications of Western Europe. Were they really? Perhaps not, but it was worth trying the comparison, saying that "the Romanian five year course was closer to a German or French diploma of a *Grande École*, than to a combined English BA and MA".

As they listened patiently to my explanations, I added the sting in the tail:

"Of course, a Romanian MA degree would be worth more than its Cambridge or Oxford equivalent".

"How come?" asked the startled Vice Chancellor.

"Quite simply: at Cambridge you only need to pay a five pound fee to automatically obtain an MA, once you are 25 and already have a Cambridge BA. By contrast, in Romania you have got to submit a dissertation and defend it in front of examiners, before you obtain your Master's degree."

"Point taken", said the Vice-Chancellor with a terse voice and a wry smile, adding some notes to the file in front of him.

My rejection was inevitable and I had asked for it out of a sheer irrational, if self-destructive streak, but somehow I felt relieved not to have been offered this job.

JOIN OUR TRADE UNION!

It was time to do some more serious job hunting and pay less attention to vague promises from has-beens.

Things looked a little bit brighter, all of a sudden, when I was short-listed again, this time by the Coal Board. I was called to Hobbart House, behind Buckingham Palace, where its HQ then stood, before they were moved to the Midlands. I was not the only Cambridge candidate to be short-listed, as my contemporaries, Bill Limond and Anton Ziolkowsky, were also called for interview.

The Coal Board, at the time, was concerned about losing revenue, as the coal seams were offset by geological faults. This caused the direction of coal extraction to be diverted, by many feet at a time: the cost of changing the gallery geometry was substantial and it was imperative that means be devised whereby such fracture zones could be predicted by mathematical means. The project was quite exciting and required as much imagination as hard maths, for which reason some heavyweights were considered and we found ourselves on the short list.

We were interviewed separately and I remembered the *sine-qua-non* condition of joining the trade union, in the case that we were accepted.

I told my interviewers that I had no desire to join a trade union. They smiled, as they thought I assumed I had been asked to join the Miner's Union, so they explained that 'this was nothing of the sort—this was the white-collar union, not the miners's, which were poles apart".

I said,

"I had realized that I would be offered membership in a professional, rather than a workers' union, but I felt that I would not like to be a member of either. I believed, as a matter of principle, in the *'free man in a free country'* ideal. I had the right of a choice and my choice was *not* to join any such organization, which one day, by the rule of a manipulated and thoughtless majority, might compel me to do things of which I disapproved."

I did not get the job.

CHILDHOOD IN A NUT-SHELL

Shell had a particular ring about its name, for 'historic' reasons—it had been present in Romania since the turn of the century. I had recognized the orange and yellow logo since my childhood and before foreign companies were nationalized in 1947.

Shell had its social club at Câmpina, where most of its executives sent their families for the summer. This resort was situated some 1200 ft above sea level, in the Carpathian foothills, before the mountains announced their presence, with the pungent scent of fir tree resin. Câmpina had a pleasant climate, typical of the hills, not too hot during the day and not too cold in the summer nights, as was the case at Sinaia, where the Royal Family had their summer residence.

After the war, during the summer of 1946, when I was only five years old, Father rented a villa at Câmpina and came regularly at weekends, in the company car, to join his family. He had several of his former university colleagues working for Shell and spending their holidays, as we did, at Câmpina. Father often took my sister and I to swim at the Shell Club. This was considered to be 'a treat' and a privilege not to be abused. I had to be 'a good boy' to be taken for my daily treat. If I was naughty and misbehaved I might risk the shame of being left behind, unable to cool off in the Shell swimming pool.

Given these circumstances, to me, as a five-year old child, Shell became synonymous with the privilege and wealth of an exclusive club.

When the Shell advert appeared in the *Sunday Times*, offering openings for university graduates with a PhD, I jumped at the opportunity and applied. Teddy was my usual referee. He was also a Director on the Board of Shell and was often whisked, in an executive jet, from Cambridge airport to some international destination, on company business.

Shell had given our Department of Geophysics at Cambridge several research grants—my chances were good. I was therefore not surprised when I was short-listed. Better still, I was considered to be of a high enough calibre not to be interviewed in London, but to be sent straight to the International HQ in The Hague.

I was happy to go to Holland again, although it was in winter. The journey from Cambridge to The Hague, via Harwich, took a whole day on the slow East Anglian train and cross channel ferry. The sky and the sea fused together in homogenous milky chickpea soup. 'Holland-by-the-Sea' was wet in a fine rain and the hotel in Scheveningen, where Shell had reserved my rooms, looked bleak and unwelcoming. I was not going to be downcast and decided to boost my morale by sitting down to a hearty meal. It was well deserved, I thought, as I had set off from Peterhouse at the crack of dawn. It was nearly nine o'clock in the evening and I had no time to freshen myself up before catching the last orders for dinner. I quickly scanned the menu that the waiter had just given me.

"Don't go away. I will tell you straight away what I would like".

"Are you Shell?" the waiter asked.

To simplify matters, I answered in the affirmative.

"In that case", came the stern warning "you cannot eat *à la carte.*"

I looked incredulously at the cheeky Dutch waiter.

I ate *à la carte*. After this dialogue, fraught with danger, I went to bed, not without promising myself to report this rather unpleasant incident to the Shell HQ, the following morning, as I felt that such initiative, by a mere waiter, would give the company a bad reputation.

HATING VAN GOGH

We were several young graduate candidates for the interview, who gathered early in the morning at Shell's International Headquarters in The Hague. The buildings had the air of a huge hospital, or hospice, rather institutionalized and pre-war in style, well kept, but dated. We were herded together in the 'catacombs' of the building, where the company photographer took two snapshots of each of us—one full front and one in profile. I resented the one in profile. The idea of two mug shots smacked of a prison camp. Did they also want our fingerprints? A shiver went down my spine. No, this was just the chill of the basement—nothing sinister about Shell, only a very odd start, very odd indeed! I shrugged off these unpleasant thoughts, which brought back memories of Communist Romania.

We were sent back to the reception hall. A blonde, middle aged secretary came to take charge of us and handed over to each of us two sheets of paper, with the two-day programme. I took the opportunity to take the blonde secretary on one side and let her know, in confidence, about the very odd behaviour of the waiter in the Scheveningen restaurant, the previous night. She looked at me intently, straight in the eye, with an expression of disbelief:

"How very extraordinary!"

"Yes, indeed, Ma'am. It gives Shell a bad reputation. These incidents should not happen, it is very bad!"

My Shell interviewers were very interested to hear about my plate tectonics research at Cambridge. This was a company which was making use of the latest theories in their exploration programme. At that particular time, I was completely oblivious of the usefulness of plate tectonics to oil exploration. My views were that the theory was still in its infancy and suffered from teething problems. It was far from being perfect.

Let them be in no doubt about it!

In typically Cambridge fashion I dismissed plate tectonics as 'rubbish'. This was of course hyperbole, as all I meant to say was that the model was not to be taken as dogma and that there was a lot of room for improvements. My interviewers took me literally and thought me at least eccentric, if not outright mad.

The company car whisked me from one end of town to the other, this time to a laboratory, where it took some five minutes to negotiate the corridors, before I reached my next interviewer. This was a small, medium built Dutchman, of middle age, who had a few oil paintings on his office walls. I thought I would make some light conversation to warm up the proceedings. To show off my knowledge on art and to please the Dutchman I made a reference to Van Gogh, whose biography *Lust for Life*, by Irving Stone, I read before I was eighteen and nearly failed my exams for not reading the recommended literature

"Who?" the Dutchman asked.

I repeated: "Van Gogh."

"Never heard of him." I said:

"Van Gogh, the Dutch Impressionist, who painted at Bourinage and died in Arles."

"Ah, you mean Van *Hooh*" he added in a guttural voice, as if to clear his throat.

I realized instantly that I must have mispronounced the Dutch painter's name, in a French manner. I felt awkward about my mistake.

"I hate him", my Dutchman added.

I felt even more awful.

He soon led the way along the labyrinthine corridors, back to the waiting car, another five minutes walk, which I thought took ages, as we did not exchange a single word, like a school master taking a naughty boy to his punishment.

On arriving at the car, I suddenly remembered that, in my flustered confusion, I had left behind in his office my Cambridge chequered country cap (very English), which I had bought, for the occasion, at 'Ryder and Amis'.

"Aghhh!" he exclaimed in disbelief, with evident sign of distress, underlined by the guttural sound produced from the back of his throat, just as he had pronounced 'Van Gogh' earlier.

"Aghhhh!" he repeated again. He turned on his heels instantly and ran for his life, as fast as he could, back to his office to fetch my cap. He returned, short of breath, to the waiting car:

"Now, you are going to be late to your next interview", he said, on handing over the lost apparel. I said with some satisfaction:

"Thank you, how very kind of you—to run."

The 'run' bit was intended to add a derisive twist, but he did not react, as he was still out of breath.

Back in the car my thoughts drifted in another direction. I felt, all of a sudden, that I was most certainly not made for this organization: this looked to me more like a military organization, where one was regimented, put in cubicles, fed on a fixed menu (not *a la carte*!) and with a convict mugshot for an ID. That is what it was all about, *and* into the bargain, with mightily awful weather and headquarters which looked like an hospice. Shame about the swimming pool though!

The remaining interviews did not matter. There was a singular similarity between the inmates of the various offices. Some sort of institutionalized face, or faceless people, I could not exactly pinpoint it, like innumerable twins:

"All these people look the same, think the same, speak the same, behave in the same manner. They all hate Van Hooh!"

By the time I reached this conclusion, I had also reached the end of my interviews, bar the last one. This was the most important person of all, the 'General Manager of International Exploration and Production', Dr Beck, a Swiss national.

THE BOMB-SHELL INTERVIEW

Dr Beck's secretary, a young spirited person, not particularly beautiful, but very pleasant, came quickly to me and said smiling:

"I am afraid Dr Beck is not here."

Although, by the time of my twentieth Shell interview, I could not give a bean if Dr Beck was there or not, upon hearing this statement (said with a flourish), I experienced a mixture of shock and curiosity. I was curious to know why Beck was not there? It was terribly rude of him to have his name on *my* list of interviews and not to turn up, as it were.

This simply couldn't happen in Shell, where everything was charted out at least two days in advance and went like clockwork, not a minute late. Besides it was a question of *image*, like the story of the waiter in the Scheveningen restaurant, with the *à la carte* menu: it could not possibly happen in Shell! It was bad, very bad indeed!

The young secretary scrutinized my face, waiting for a reaction, whilst my thoughts must have portrayed me as an absent minded, vague fellow. I answered almost like an automaton, who, for two days, had been regimented in spurts of half-hourly interviews, with square heads, in white collars and neatly trimmed hairstyles:

"Dr Beck is not here? Never mind, I can wait: my ferry from Scheveningen does not leave until late."

I looked at the décor around me and it was quite different from the other offices and very inviting for a nap. I was exhausted and I felt this was a welcome opportunity during which I could sit comfortably in the plush leather armchairs in Dr Beck's antechamber and recharge my batteries, whilst he went about his business. Even the sofas in the wide corridors seemed to me OK to lounge on, if I was not wanted in the antechamber.

Here everything pointed to power and influence, with every distinct sign of affluence marked with military precision. Dr Beck was, after all, at the top of the pyramid in Exploration *and* Production of a powerful multinational company. He had his own secretary, a larger office than anybody else, with beautiful views from the top floor, a partner's desk the size of a king's bed, a huge antechamber with a battery of telephones, to compete with those of the White House or the Kremlin, a new Persian rug.

"How many knots per square inch?" I wondered. "Must be a lot, all paid in gallons of crude oil, extracted by Shell engineers, through Shell pipelines, transported by Shell tankers into Shell ports, to be refined by Shell refineries and sent to the Shell forecourt pumps, to pay for the swimming pool in Romania *and* for Dr Beck's Persian rug, of course."

It was all a neat little story, which fell into place. The secretary disappeared for a moment and returned to let me know, in an unequivocal voice:

"I am terribly sorry, I checked everywhere and I cannot find him. He must be in a meeting."

" Not to worry, I have two hours before the ferry leaves, plenty of time and my overnight bag is packed. I can wait for Dr Beck. It is important that he sees me. He is on my list."

I showed her the list where Dr Beck's very name was at the bottom. It said:

"4.30 pm: Dr A Beck, General Manager, International Exploration and Production, Europe, Africa, Middle East, Asia, Far East, North and South Americas."

These titles were like the jewels in the crown, like the series of titles that Grand Dukes or Palatine Princes added to their names.

"Oh, my God", now I noticed: "They forgot to add to his title all these other places: they forgot about Central America, Australia and Antarctica! Did Beck know that his empire was curtailed? Maybe he needed a Cambridge man, with a PhD in plate tectonics, to bring the unforgivable omission to his attention, as nobody else had the courage to do it!"

It was only years later that a friend from Shell confirmed that if Beck was 'unavailable', this simply was a sign that the candidate had got the chop.

I never even noticed that the soft-footed secretary disappeared for a moment and came back, for the third time, with a very different message:

"Dr Beck is here. He wants to see you."

"Good!"

I straightened my tie to prepare myself for the great moment and fretted with juvenile anticipation. The blonde secretary led the way and announced me in a mellifluous voice:

"Mr Constantin Roman."

She closed the padded door behind me. Now I was facing the great man himself!

He did after all have time to see me! How good of him!

I looked at him, with great bovine eyes, waiting for God to make a pronouncement. He did it suddenly and unexpectedly, with a snarl:

" You are conceited!"

What? Did I hear right? Did he say 'conceited'? Surely, I must have misheard him! He must have said something completely different: this was hardly the way to start an interview, hardly the way to treat a young foreign guest: I did nothing wrong, just suggested that I could wait for him. No need to insult me for it. How dare he insult me? Only a KGB colonel would have taken such liberties with his victims. True I had my picture taken full front and in profile, in the basement of Shell, but this was no carte blanche for insults.

"You are conceited", Dr Beck thundered again, thinking that I may not have heard the first time round, that I may have obliterated the sound, filtered it out, for convenience's sake. Who wanted to hear unpleasant remarks? Not even the deaf, as they could lip read.

Now for certain, I heard it right, the second time, the terrible indictment, like the pronouncement of the Inquisitors who sent the heretics to the pyre. I felt like a heretic. I was not, thank God, like any of them. I was my own man and proud of it too! I straightened my spine and moved the head up, defiantly. I was ready to return the compliment:

"So are you!" I answered.

I could not care less as to the number of continents the omnipotent Beck had in his empire—I turned on my heals and left Beck's office.

I was quite relieved at the thought that I had missed nothing, that I would not regret an iota not being offered a job there. Or did I? Were there any lingering regrets? Maybe the swimming pool of my childhood could never be forgotten.

What long-lasting damage such memories could have on one's psyche! Although I tried to dismiss this incident from my mind, I felt a prisoner of my childhood in Romania.

"How ridiculous! How *very* ridiculous!"

ALIEN OR A LION?

Long before I knew that 'alien' was synonymous with 'foreigner' or 'foreign' this was an abstract notion in my passive vocabulary, with no particular meaning attached to it.

It was only when I passed through Heathrow airport, on my first visit to England, that I had seen the word 'alien', which must have meant 'foreign' because I could not gain admission through the gate showing the sign 'British nationals'.

Still, I was not completely convinced that 'alien' was interchangeable with 'foreign', because one would have had an 'Alien Secretary' instead of a 'Foreign Secretary' in the Government and an 'Alien Office' instead of a 'Foreign Office'. Mind you, what went on behind closed doors in the Foreign Office was pretty *alien* to nearly everybody. This would explain Margaret Thatcher's profound mistrust of the latter, or Chancellor's Kenneth Clarke's mistrust of the Foreign Office having the greatest demand on its budget for entertainment with food and drink, which is alien to the British public.

Hence the rather derogatory nuance of the rude 'alien', as opposed to the more urbane, yet still mistrusted 'foreigner'. To make this point clearer, the best of the latest such rude usage for the word 'alien' was coined during the Second World War, when scores of foreigners were equated to 'alien enemies' and transported to isolated enclosures.

No need today for such semantic excuses for putting *aliens* behind bars, the fear of the unknown sufficing as a motive for persecution. In any event, for this particular purpose, when they entered Britain, *aliens* would have to be 'processed' through certain channels, like processing jute fibre before it could be woven into doormats.

To this *alien* the great zest of its appellation would reside in its original mispronunciation of 'a lion', which brought about associations with certain proud creatures of the wild, which were deemed to be kings of the *Animal Kingdom*, a lion-king, that is. In a subconscious way being a lion made me, *inter alia*, a king, as in Rudyard Kipling's stories of my childhood.

During my interview at the Home Office I made abundantly clear such feelings to the startled officer, whom I had reassured that:

"Britain would like to have 'a lion', such as I, to settle here for good, but I had no such plans, as I wanted to settle elsewhere, once my studies were finished."

By this I meant, immodestly, that should I settle here, I would be an asset to this country and Britain would be happier for it.

One way of looking at the value of *aliens*, which was never tried before, or, rather took a long time to be acknowledged, was the beneficial effect of the Norman invasion of Britain, or the 'Glorious Revolution' of William of Orange, or the arrival of the Huguenots, not to mention Emperor Hadrian, who introduced central heating to Bath. Some two thousand years on, the Polish pilots too did their bit in 'the Battle of Britain'.

In any event, no matter what merits aliens might have had in the past, this particular alien, on arriving in Britain, had no intention whatsoever of settling here. He merely asked for leave to be allowed to stay for his studies, whilst he retained his Romanian nationality and travelled on a Romanian passport. For this reason, when his knowledge of English became sufficiently refined to understand that the word *alien* was offensive, he insisted at being referred to as a 'visitor' to this country and asked for a 'visitor's' visa.

TRIBAL CALL

In a tribal society, even a visitor could be regarded with suspicion: what if he came to dinner and stayed the night? What if he came for a weekend and overstayed his welcome, bleeding the drinks cabinet dry?

This suspicion presumes that one's own home is more desirable than the visitor's own home. For example, you might have a more attractive wife whom the visitor might wish to seduce, like those villainous Vikings who came to these shores to inseminate the English fishermen's wives.

Or, in more modern times, making political capital by manipulating the British public into believing that they had the 'best social care system in the world', the envy of other nations. It stood to reason that as a consequence of its uniquely desirable status, *alien* subjects of other nations would flock to our ports and airports to use and abuse our system: Italian girls getting free abortions on the National Health, Irish navvies abusing unemployment benefit and Caribbean black immigrants jumping the housing queue: enough to make one's blood curdle against these repugnant immigrants.

VISA TO ECCENTRICITY

It was considered at least eccentric, if not outright silly of me, to carry on travelling on a Romanian passport while I lived in the West. On one hand I was treated with greatest suspicion by both officials and ordinary denizens, and on the other, it put endless bureaucratic difficulties in my way over single re-entry visas, multi-entry visas (in the case of Britain), too frequent entry visas in any one year (in the case of France), or any visas at all, in the case of Canada.

"Why not give it up?"

"Is it repugnant to be the citizen of a Communist country?"

"Why not ask for political asylum and get on with your life, making your existence simpler, by travelling on a civilized passport, like a British passport?"

The answer was very simple: nobody would offer me a British passport on a silver platter, as it required an incubation period of at least five years of residence, before one could even apply for nationality.

Secondly, applying for political asylum several years after entering Britain could not be credible and was ruled out for other reasons.

Thirdly, for what it was worth, my Romanian passport was all I had and it was still valid.

Last, but not least, to me it was far more important that this passport identified me with the country where I had my roots and where I was born, than the fact that its régime, at the time, was of a Communist hue. It was not my fault that Romania was Communist and I did not identify myself with that ideology. Too bad if simpletons thought otherwise.

In the meantime the inordinate amount of time spent in endless applications for my scientific travels and conferences and the endless agony over the waiting game which this involved, added to the frustrations, not only mine, but those of the people around me in Peterhouse and the University whose function was to look after my welfare.

I was still under the illusion of being a 'free man in a free country', but there was no such thing: I could not be treated as equal to the citizens of the 'Free West', simply because of the shackles imposed on me, as a Romanian citizen, through an iniquitous division of Europe after the war.

EXTRA PASSPORT PAGES

Although I was the recipient of all these painful travel constraints, I did my best to brush them aside and travelled whenever I could in Western Europe, on my Romanian passport. Most of the time this was for professional reasons, either to attend conferences, give papers or gather data for my dissertation.

Usually, at the end of any such visit abroad, I would extend my stay for personal enjoyment, visiting new cities, monuments and museums. My thirst for understanding the relationships between different styles of architecture in different countries, or the influences of one school of painting on another, was difficult to quench, as I knew it was such a vast and diverse subject.

I had to be selective, knowing that I usually travelled on a 'shoe string' budget, but it was tremendous fun and I was overjoyed at this 'newly found freedom'. It was more of a 'freedom of the spirit', than real freedom, but infinitely more than I would have been able to get within the confines of Communist Europe. *Vive la différence!*

So far as the Romanians were concerned, I was not supposed to travel to countries other than the one for which I had originally asked the passport: Great Britain, for which I had wrenched my famous *visa de călătorie in Anglia*. It was typical of the Party commissars not to be able to make a distinction between 'Marea Britanie' (GB) and 'Anglia' (England). So, why should I go, hat in hand, asking permission to go elsewhere? My passport did not say that it was valid only for travel to specific countries, only my 're-entry visa to Romania' said so. Yes, believe it, or not, Romanians had to obtain a visa to re-enter their *own* country, very much like *aliens* in GB needed a re-entry visa.

This simply meant that all Romanian citizens who travelled abroad were automatically equated to 'aliens', who needed a re-entry visa, just in case they had to be put in quarantine, for having caught abroad the political equivalent of the rabies. Unless they were deemed, by the authorities at home, to be 200% 'reliable', Romanians who travelled abroad were not trusted by the Ceauşescu régime, lest they had been contaminated by some 'funny' ideas in the free West. I fell into the latter category and the comrades from the Romanian Embassy, in Palace Green, regarded me with automatic suspicion:

The comrades from the Romanian Embassy in London were slow in reacting to my Tutor's plea for more pages in my passport. This was an unusual request. They had to get orders from 'the Centre'. But first and foremost, they ignored the letters, thinking that Peterhouse would give up in despair: the tactic best used was 'to defeat one's enemy by sheer inaction: he will eventually get tired and go away'.

We did not! I kept prodding my Tutor, Dr Lovatt, to write again and again. Persistence eventually paid off: we had to ask the Vice Chancellor of Cambridge University to request more pages in Constantin Roman's passport, 'in order to be allowed to travel to collect crucial data for his PhD thesis', before the Romanians relented and eventually sent a further eight little passport pages. At long last, the Romanian Embassy in London understood that it was a *casus belli* and that it was better to relent, but only after having made us wait for six months and made us go through endless correspondence. To what end, I wondered? It was absurd! They did their reputation more damage than good, but it did my case a lot of good, by gaining sympathy for my cause, from all round, for having my basic rights abused in this manner.

EMBASSY COMMISSAR

The episode of the passport's extra pages focused the attention of the Romanian Foreign Office in Bucharest on the presence of a 'troublesome Romanian student' in Cambridge.

It is an established fact, in any warfare, that ambush is the best tactic to cause maximum damage, by taking one's enemy by surprise and securing victory with the minimum of effort. The comrade from the Romanian Embassy in London, whose task was to 'bring me to heel', had planned to do just that: he came unannounced, parking his chauffeur-driven limo just inches from the hostel's front door. Doubtless the comrade was very good at setting up traps for wild game in the Carpathian mountains, but here we were in the middle of Cambridge, opposite the porter's lodge of Peterhouse. His brief was maybe to bundle me senseless in the back of his limo and take me to London, where some well-trained embassy contortionist would have extracted from me anything they wanted.

As it turned out, I managed, through sheer foresight, to extricate myself from the confines of the hostel, and lure my stalker into the open surroundings of the Graduate Centre nearby, where scores of friends were having lunch. I also made sure that they knew what odd companion I had with me. The comrade's excursion to town did not go according to the pre-arranged plan: he tried to convince me to apply for a special kind of passport for 'Romanians resident abroad', or simply to renounce my Romanian citizenship.

I told Comrade Spătaru, Second Secretary, in no uncertain terms, that I was proud of my origins and I had no intention of making a present of my nationality to anybody: it was an unalienable right of birth.

As for a passport for a 'Romanian resident abroad', that was no good to me, as I intended to return to my home country, as soon as I had obtained my PhD. All I required from the Embassy were more pages in my present, valid passport and an up-to-date re-entry visa to Romania.

LESS COMRADELY COMRADE

The comrade left empty handed, but a new one was soon to come—also unannounced—except that on this occasion he could not park his limo in front of the hostel door as by then I had moved to married quarters behind the hostel in Cosin Court. As this building had no direct access to the street, only pedestrian access, here the avenging comrade was left pounding the door of my flat for ages, completely oblivious of the fact that I was watching him, with great interest, from a top floor terrace.

I reported the incident to the Peterhouse porter and he suggested, if this happened again in future, I should not hesitate to call the police. Doubtless the comrade might have claimed diplomatic immunity. What kind of diplomatic immunity would one require before such beef would try to break down the entrance door of some student residence in Cambridge? Such talent was in greater need at London Zoo, for the brutish keepers of the rhinoceros, rather than in the embassy employ of a European nation, be it Communist or not.

I soon had visions of Eugène Ionesco's plays in Paris, which were banned in Romania because of their political innuendoes.

A SUDDEN INTRODUCTION

The banging on my door in Cambridge by the brutish Romanian emissary brought home the stark reality, which I was unready or unwilling to face: there was no turning back, there was no 'U'-turn in a one-way street.

Did not Lipatti, the Romanian ambassador to UNESCO in Paris, say in 1968 that 'doing a PhD in the West was a political option'? How right he was! It had taken me all these years to reach the same conclusion as this Securitate ogre!

Maybe, unwittingly or subconsciously, I took a *political option*, which now denied me the ability to return to my home country, be it a home-prison, where the freedom which I had enjoyed to date might be considered an infectious virus, which no amount of 're-education' could ever cure. The only known cure for such a virus was to burn out the patient: I had no courage for such treatment and from this moment on, I felt that a major turning point took place in my thinking: I decided actively to seek permission of abode in the United Kingdom.

I figured out that even if I did not receive a job, at least I could use the UK as a base from which to travel to whichever country would offer me work. I had to have a home: the ideal of any self-respecting, civilized person was to settle down and put roots.

By now, I had lived in the UK for nearly four years, the longest period ever spent in any country, other than Romania. Here I had studied and absorbed avidly the culture surrounding me. Here I had cried, laughed, made friends, I travelled to the cathedral cities and piously read the centuries-old scrolls. Here I had shown tourists around my beloved corners of Cambridge. Here I had hated and loved. It was not until I had seen again the white cliffs of Dover on the horizon, on one of my return trips to England, that I was surprised at my unexpected deep exhilaration and I realized that England had grown on me: I had fallen in love with the old girl.

There was more to it than anglophilia in this realization: here I had met Roxana and we both felt that we wanted to make our home together in England. I knew that, before I married her, I would have to obtain permission to stay, simply on the merits of my own case, rather than by being married to an English woman.

But how was I to start in this endeavour? I knew that so long as I was a student, or rather because of it, my application for permission to stay in the UK would be automatically blocked by Home Office regulations. I was further advised that the political asylum route was 'not a runner' and even if it were, I would not have wished to put my Romanian family at risk. The only way in which I could convince the British authorities that I had good grounds for being allowed to stay, was to prove that on my return to Romania I would suffer persecution. True, everybody might dream up such a story, but I had to prove it. It was like a move on a chess board, where the game has to be played by the rules, or like a court case, where it would not be sufficient to be right, one would have to prove it, by making a good legal case for it. How would I be able to make a good case? Not on my own! I needed somebody credible, who was familiar with the realities of Communism and, at the same time, a respected figure here, willing to champion my interests to the British authorities, to espouse my cause as a good cause. This was a tall order to fulfil. Who would do the honours as the white knight in shining armour, ready to slay the bureaucratic dragon?

I had no need to look very far afield, as only a few doors away from Peterhouse lived the new Master of Corpus.

Corpus Christi College had two entrances: the main entrance, in Trumpington Street, through the 19th century Gothic revival gate, leading to the main court, and another by St Bene't's Church, in Bene't Street, opposite 'The Eagle'. This latter entrance was through a modest 14th century Gothic archway, leading to the Old Court, the only complete example in Cambridge of what a Medieval College might have looked like. It is in this very court that the poet Christopher Marlowe lived between 1581 and 1587, the year when *Tamburlaine* was finished, which inaugurated blank verse.

The Old Court witnessed the 'Black Death', medieval town riots and the plague and therefore has its share of ghosts, amongst whom lingered the lover of the Master's daughter, of the 17th century. He was disturbed during a clandestine visit and sought refuge in a cupboard, where he suffocated. The College has the unusual disposition of a Hall on the first floor, reached by a flight of steps through the passageway linking the Old and the New Court.

Walking through Cambridge Colleges was like walking through a living museum, where a lot of new philosophical thinking and scientific discoveries were made in the shadow of old buildings, full of historical meaning. It was a delight to the eye and to the spirit to search one's surroundings, to discover new corners which unsuspectedly displayed art treasures, whether a sculpture, a painting, an old stained glass window, or an ancient manuscript. The custodians of these treasures were also more often than not very interesting people, who would stimulate one's mind and imagination.

One such example was the Master of Corpus, a retired diplomat and Oxford-educated historian, Sir Duncan Wilson. It was rather unusual for a Cambridge College to invite Oxford men to become College Masters and the most recent examples were not a great success, either at Peterhouse or at Corpus. Duncan Wilson's last post, before he retired, was that of British Ambassador to Moscow.

As an academic and a humanist, he would understand my anxieties. As a former ambassador to the Soviet Union, nobody would be better placed than him in understanding the viciousness of the system within which one false step incurred severe punishment. Better still, during his time in Moscow, Sir Duncan's daughter, Lisa, went to the celebrated Conservatoire to study cello and there she met and subsequently married the Romanian pianist Radu Lupu, who was popular in Britain as the winner of the Leeds Pianoforte International Competition, in 1969. As the father-in-law of a Romanian now living in the West, the former Ambassador would be personally acquainted with the Romanian brand of political practice.

I had several friends who knew Duncan Wilson, but I decided that I should try the direct approach and make a case for myself, without any prior introduction. I did not make an appointment either—I just knocked at the door of the Master's Lodge, saying that I was a Romanian student in Cambridge and asked to see the Master. As Lady Wilson had shown me in, I entered the Master's study with a determined look on my face and said:

"Sir Duncan, may I introduce myself?"

In retrospect this sequence still makes me smile, as it had a shade of adventure and pioneering spirit, which prefigured the inevitable answer:

"Mr Constantin Roman, I presume?"

IMPROBABLE MATCH

It was the beginning of a long and unexpected complicity: nothing could have been more improbable than the meeting of our two minds: he was thirty years my senior and a seasoned diplomat. I was young and a very undiplomatic person. He was educated at Winchester and Oxford—I was educated at a Romanian State school and at Cambridge. He was a humanist—I was a scientist. He was tall, with an Anglo-Saxon complexion, I was of moderate height, with Mediterranean looks. He spoke the best Queen's English—I professed English with a pronounced Latin accent. He lived all his life in a democratic society—I lived nearly all my life in an oppressed society. He was a College Master at Corpus—I was a College student at Peterhouse.

It was very surprising indeed that, in spite of all these pronounced differences which separated us, the chemistry was right from the beginning. Maybe this was because we were both idealists, ready to force the world to fit our preconceptions of rights and wrongs, both feeling very strongly about injustice, being bullish and ready to be a good sport, taking on a challenge at any time and, above all, taking the bull by the horns and making it kneel.

All these common qualities made us into a formidable working team, whilst the differences gave our tandem strength through complementarity.

There was no need to explain at length the intricacies of my case, as Duncan Wilson was trained to absorb the essential very quickly, to see clearly the whole picture in perspective. It was a great pleasure to brief him, as in the past it had been a constant pain to warm reluctant hacks up to my cause, who tended to remain lukewarm on the battlefield. To the latter group I had to give constant kicks, whilst to Duncan making a statement was enough to inspire him.

Was it just plain sailing? A stroke of luck or genius? Was it God's miracle? I could not know! The point was that we understood each other perfectly and, from then on, we would work hand in glove, until the problem was solved. Thereafter, we remained friends, although on leaving Cambridge I went abroad and Duncan retired to Islay, in the Western Isles of Scotland.

The crux of the matter to be solved was proving to the Home Office that on return to Romania I would be persecuted and as a corollary that I should be given leave to stay indefinitely in the UK.

A DIPLOMAT'S STRATEGY

"Have you got any idea?" asked the former Ambassador to Moscow, taxing this student's resources to the full.

Here is where our complementarity gave us strength: first I would not take 'no' for granted: had some 'wet' from the Foreign Office gone back on his word, that would be too bad! Sadly, but not surprisingly, such disappointment had already occurred, during Duncan's first attempt in enlisting the Foreign Office's help on my behalf, in order to demonstrate to the Home Office that my fear of re-entering Romania was genuine and that the risks involved were very real indeed. Such a statement would have sufficed in solving my problem, but the FO having given such an undertaking, had gone back on its word! Even though he was retired, Duncan was too loyal a civil servant to share his disappointment with me, but such ungentlemanly behaviour by one of his former colleagues went against the philosophy of rectitude of this Wickhamist! My analysis of the situation, after this initial setback, was rather different from Duncan's: one had to try somebody else and, at the same time, give stick. Were they too polite to intercede? Then push them: they would love every minute of it—after all, there was enjoyment in brinkmanship. Show them to the abyss, then pull them back, by the scruff of their neck and they will be grateful to you for rescuing them from sure death.

What should we do?

I said: "There is not much else to do than use a good lawyer to take up the case."

"Have you got anybody in mind?"

"Yes, I have."

"Who is it?"

"Lord Goodman."

"Leave it to me" he said quietly "I will write to Goodman."

Within days I received by post a copy of the letter Duncan wrote to Goodman, accompanied by a compliment slip from the Master's Lodge, with a brief hand written note:

"I hope this is to your satisfaction, Yours Sincerely, Duncan Wilson."

In his early diplomatic career Duncan Wilson was at the British Embassy in Belgrade, and most certainly figured, under an assumed name, as a character in Durrell's *Esprit de Corps*, by which Duncan let me understand that he was certainly 'not amused'. Duncan Wilson was busy at Cambridge finishing a book on *Tito's Yugoslavia*, in which he had the foresight to predict the political turmoil that would tear this country apart after the dictator's demise.

Lady Wilson was not going to be outdone. She was a blue stocking and did her own thing: she had learned Russian and produced a most impressive bilingual English–Russian dictionary. This, evidently, had its domestic disadvantages, as the meals served in the Master's Lodge, which I attended on more than one occasion, were very frugal, usually composed of a soup and cheese. Was it a conscious effort to keep in trim, at a certain age and 'eat healthy'? Or, was it rather a welcome holiday from the elaborate diplomatic and College dinners, which must have been a strain on the liver? More likely, I would have surmised, in an uncompromising mood, Lady Wilson's own professional pursuits were in conflict with her domestic duties in the kitchen. How rude of me to think so, when it was a privilege to be asked to their informal table, in the Master's Lodge at Corpus. But then, may I be excused, as all East Europeans deem food as absolutely essential, through the memory of many generations of underfed peasant ancestors? Besides, I was much younger, as I could have been their son, and used up a lot of calories playing squash and cycling in Cambridge. Still, on reflection, it was the company and not the food that mattered!

Lady Wilson was kind, practical and matter-of-fact. Being an East European chauvinist I paid more attention to her husband.

Duncan Wilson had an innate courtesy and a face always lit by a benevolent, engaging smile, always inviting one to open up to him. He was tall and erect on his large body frame and his presence made a strong impression. His conversation was warm and put his interlocutor at ease. He talked quietly and methodically with the logic typical of an academic. A very happy combination of an old school well-bred diplomat and erudite academic. This put him in a special category, of a kind which was unusual in Cambridge and which probably brought him into direct conflict with the college Fellows. Not an unusual occurrence in Cambridge, but a rather unhappy circumstance for the Wilsons, who deserved a more peaceful life. This, eventually, they have achieved in their remote retirement, on the island of Islay, perhaps a refuge 'away from it all', but an unusually isolated place for an active retirement.

HUNGARIAN SKEWER

There is nothing more lethal about grilling a Romanian, then to put him in the hands of a Hungarian: the two nations are known not to get on terribly well with each other, to put it mildly. Maybe Duncan knew this and he specifically engineered this provocation, to make sure that the worst would come of my encounter with Lajos Lederer, a columnist at the *Observer*. It was like asking the Devil to vet the souls of the righteous at the gates of Paradise. Lederer proved to be a meticulous *agent provocateur*, a real torturer, and I presume he would have made an effective interrogator for many a secret service in Central and Eastern Europe. I personally had no choice but to meet Lederer, as Duncan Wilson said casually:

"Why don't you go and see Lederer in London?"

I assumed that it was another initiative at opening a door on the labyrinthine corridors to the residence permit. Why not? If Duncan says so there must be a good reason for it: I will not say no, I'll perform any sacrifice, even talk to a Hungarian!

This was to be another of those grilling sessions of the kind I had lived through earlier. I was used to nearly everybody asking nosy, even very personal questions about my past and my reasons for being here, to the point that I no longer found these questions rude, or unusual: I assume I became quite *blasé* about it.

My attitude was one of openness, as I had nothing to hide. There was, however, a human limit to the degree of being pilloried with questions, as if I was a criminal: how demeaning to have to carry on performing this kind of moral striptease.

Whatever the case may have been, Lederer reported back to Duncan Wilson that I was the 'genuine' article and not a fake. This gave Duncan enough confidence to write to Lord Goodman and ask for help on my behalf.

UNNECESSARY INHIBITIONS

One morning, soon after my Hungarian interrogator grilled me, I received a short note from Goodman's secretary, asking me to ring the office to make an appointment to see Lord Goodman.

At the time of this invitation, Lord Goodman made the headlines in nearly every day's newspaper. He was one of England's most reputable lawyers, adviser to Harold Wilson the British Prime Minister, and Government negotiator on Rhodesia at a very important turning point, when Ian Smith had declared unilateral independence by a white minority Government. Apart from his political and legal activities, Goodman held many honorary and executive positions, being, amongst others, Chairman of the Arts Council and Director of the *Observer*. His name would conjure in the British public as much reverence as the best goalkeeper of the top football team in the national league, or that of the best cricketer, as Goodman played cricket by the rules and always won the match. He was a formidable adversary and an indomitable ally to have.

Arnold Goodman was born in a Jewish immigrant family from Estonia, who came to England at the end of the 19th century, following the pogroms in the Russian Empire. Most Jewish immigrants worked hard and gained positions of influence in industry, banking and politics, which made Harold Macmillan say that there were 'more old Estonians than old Etonians' in the British Parliament. This did not mean that the Goodman family made it straight to top of the ladder, far from it.

Goodman made his own way up the greasy pole, in a more conventional manner: he was educated at University College London, had a successful solicitor's practice and was made a life peer in 1965 for his services to the Nation, as Baron Goodman of the City of Westminster. In this latter capacity, he took his seat in the House of Lords.

When I met Arnold Goodman, he had further honours bestowed upon him by the Establishment, this time the more prestigious CH. This did not make him any different, as his habit was affable and modest, whilst his eyes, with bushy eyebrows, scrutinized you with a penetrating look. He had a large square face, with a treble chin and strong features, the very face which a cartoonist would like to lampoon. His reputation was as good as his name, as he was a generous, goodly man, not unknown, during his days as Master of University College Oxford, to give legal advice to many a student, completely free of charge. I fell into the latter category.

The weight of Goodman's reputation was such that, entering his large austere office made me uncharacteristically silent, a nervous lump in my throat obstructing my speech. I was supposed to be there to brief him about the scope of my visit and I had not rehearsed the agenda and objective at all. I fell, instead, into some sort of idle stupor. He tried to put me at ease from my evident discomfort and asked a few questions. I answered in an economic way, in monosyllabic utterances. I realized too well that, for me, the great moment of my life had come and that we were not making progress. So, I plucked up enough courage to justify my shyness, by admitting that I was in awe of my interlocutor's reputation, as a result of which I found it hard to talk.

The old man dismissed my suggestion with characteristic common sense:

"Well, forget who I am and say what you want to say."

It did not require many sentences for Goodman to see the core of the problem. He called in his secretary and dictated a letter in front of me, addressed to David Lane, then Secretary of State at the Home Office, in charge of Immigration.

"Let us see what comes out of it", he said quietly.

'INELIGIBLE FOR ASYLUM'

In the meantime other guns were lined up in the direction of the Home Office: this time William Francis Deedes, MP (later Lord Deedes of Aldington) and Editor of the *Daily Telegraph*, was brought into the fray, with David Floyd as go-between. Bill Deedes advised 'strongly against political asylum, which was not a runner'. He also advised 'making some concessions to bureaucratic resistance'.

Clearly, the Home Office was not going to be hurried on the way, although to me time was of the essence. Goodman was not deterred by this first inconclusive answer, he always sent copies of correspondence to me at Peterhouse and wrote to Wilson, at Corpus Christi, reassuring him that he would continue to pursue the matter until the case became crystal clear.

In Goodman speak, as I was to realize later, this meant, until the case was resolved to *his* satisfaction.

A WELL-WRITTEN LETTER

Four years on, after my arrival in Britain and one more year before my Romanian passport would expire, I still had a long way to go before all the conditions put in front of me could be met.

The chess game was proceeding, one move at a time, tediously but surely, and following each move it was necessary to consolidate it, by explaining the implications of the new situation and declaring the next short term objective. One such letter, which I wrote summarizing the situation to David Lane, the Home Office minister, a copy of which I sent to Goodman, brought his Lordship's praise for 'being well written'. This boosted my morale in the middle of a difficult battle. I was quite chuffed, as a foreign student geophysicist, to receive encouragement about the style and structure of my letter from Britain's most distinguished lawyer: I felt I was getting better and it proved, in the long term, an excellent discipline in dealing with future complex legal matters.

FIRING ALL GUNS

There was no time for rejoicing over small victories, so long as the main objective of being allowed to stay in the UK and be given permission to work was not reached.

It would have been too naive of me to hope that all those friends and acquaintances whom I had approached could humanly commit all their energy and time to my cause: it was asking too much of them and it would have been completely unrealistic. The best I could hope for was to persuade as many people as possible to write at least one letter on my behalf. It did not matter who did so (though I took care to harness to my cause some of the most distinguished names in Cambridge), but the spread of these 'guns', which I would line up to fire all at the same time, was equally important. There were Professors and lecturers, Tutors and Vice Chancellors, Masters of Colleges and Departmental Secretaries, scientists and arts people, Fellows of the Royal Society and simple citizens, 'noble Lords' and commoners. As I could not drop an atom bomb, I tried at least to fire an indomitable artillery, with some very fine gun fire. The Roman rollercoaster was on the move, polite, but sure in the 'art of the possible', with all the game above board and no foul play, with the eyes firmly on the rules, the loopholes, the discretionary powers, as well as the democratic, unwritten principles of political pressure.

The latter aspect would now be called 'lobbying' and one would call in highly paid specialists to do it. I had to do it all by myself, to the great amazement and awe of my long-suffering Peterhouse Tutor. There was a 'silver lining' to this cloud, as, in the process of obtaining all the help I could get, I had the opportunity of meeting and corresponding with people I would not have met otherwise. Some of these early supporters became lifelong friends. Everybody was aware of everybody else and a 'current of opinion' was created, which put the contenders at ease, as they felt more confident about the good cause in question (e.g. my cause) if they knew the names of other distinguished supporters.

This made my correspondents address the opening line to the Home office Minister with lines, such as Teddy's:

"Dear David, No doubt the name of Constantin Roman will be engraved on your heart..."

Indeed, it was engraved on everybody's heart and I suppose that the Home Office scribes had hardly ever been so busy.

MINISTER'S TRESPASS

Not all the letters were aimed at David Lane: some were addressed to the Home Secretary, Robert Carr, later to become Lord Carr of Hadley. Looking in *Who's Who*, I realized that Robert Carr was an undergraduate at Gonville and Caius and I thought it most appropriate to approach him via Joseph Needham, the Master of his old College.

In my early years at Cambridge, I had no particular reason to meet Joseph Needham. I knew the College well. It had an early example of a renaissance gate, designed by its second founder Dr Caius, who studied medicine at Padua and who was physician to Edward VI. Gonville and Caius had a long tradition in medicine and counted amongst its alumni William Harvey, the discoverer of blood circulation. It is the only College, which had in its statute books medical rules for the admission of students:

'The College should admit no person who was deaf, dumb, deformed, lame, chronic invalid, or Welshman...'

Thankfully, the Race Relations Board and the Equal Opportunities Act did not exist in the 16th century, otherwise the College might have been in trouble for lack of political correctness.

The introduction to Robert Carr was not the only reason I had to see Needham. By the end of my second year of research on plate tectonics, I started to look at the earthquakes of Central Asia, which included Tibet, Sinkiang and other provinces of China. I had at hand a good database on recent earthquakes, but I needed some more information on Chinese historical earthquakes, as well as some brief on the history of seismology in China. If anybody knew anything about this latter subject it would be Joseph Needham, Fellow of the Royal Society and Master of Caius—he had worked for decades on a scholarly seven-volume series on the 'Science and Civilization in China' and had already published several tomes on this vast subject. He had, at hand in College, several Chinese researchers. Needham taught himself Chinese whilst at Cambridge and subsequently participated in a scientific mission, to the province of Chungking, in the early 1940s. Some of the help I actually needed was with translations from Chinese scientific articles.

I had asked Sir Edward Bullard if he knew Needham. He wrote the following letter:

"Dear Joseph (a mere compromise between Dear Jo and Dear Master), There is a Romanian student in this department who is working on the earthquakes of Sinkiang and Tibet. Could he call on you sometime? He is a man of wide interests and would like to meet you. Yours sincerely, Edward Bullard."

To this Needham answered by writing also in long hand on the same paper Teddy had used:

"Dear Sir Teddy (the retort formal-familiar I hope) Of Course. Just let me have his name and dignities and I will ask him round. I have various registers of catastrophes but am not sure whether Sinkiang is included in all of them (I hope he reads Chinese). Ever yours, Joseph."

The first impression I had of the old man, when crossing the threshold of his rooms at Caius, K1 and K2, was that of a great scholar of immense humility. He had a large body frame, slightly stooped and a thatch of white hair, almost of a boyish rebelliousness. His study was full to the brim with manuscripts and books, amongst which he was known to routinely spend sixteen hours a day. A diminutive Chinese lady, Gwey-Djen Lu, who was his assistant, hesitated a smile from behind a pile of documents.

Needham put me in touch with a native Chinese in Cambridge, a student from King's, who would be able to help. At the same time, he also put my name on his social list and so I was invited on several occasions to the Master's Lodge cocktail parties, which were always crammed with interesting people of various nationalities from the worlds of art and science. Joseph and Dorothy Needham were at home in both environments of art and science, for Joseph Needham was to become the first scientist and FRS to be awarded membership of the British Academy. The Needhams were a unique couple inasmuch that both of them, husband and wife, were Fellows of the Royal Society, both carrying out independent research in biochemistry.

Needham had put at my disposal his documentation on Chinese earthquakes, which I needed for an introductory chapter on historical earthquakes. Eventually I acquainted Needham with my residence and visa problems and he volunteered to write to Robert Carr:

"You know? He was a student here."

Of course I knew. This gave the opportunity to Joseph Needham, Master of Gonville and Caius, to reprimand the Home Secretary for trespassing in the garden of the Master's Lodge. Joseph Needham's intervention brought my saga a little closer to its end. For the first time it was clearly stated that should I demonstrate both that the Romanians were uncooperative and at the same time that I had got an offer of a job, then I would be allowed to stay on an annual basis. This facility would be renewable each year and at the end of such a four-year period I would obtain permanent residence status. Robert Carr went on to answer his College Master's charge of trespass, which he did in long hand, after the paragraph dealing with 'official' business:

"I am afraid on the first occasion I was led in through your garden, because the police had rumours of some potential demonstrators. I apologize for trespassing in this way."

Would such niceties have been discussed in any other country in correspondence between Master and former student, now a powerful Minister? I doubt it, even by Western standards, let alone by the non-existent standards of the Communist world. This was in March 1973 and the potential demonstrators, referred to by Robert Carr, had created havoc at Cambridge in the early seventies, having set fire to the Garden House Hotel. The Left was quite vociferous and violent, Government Ministers often being the focus of such attacks. Why did not the British Left export such iconoclastic practices to Soviet Russia? There, at least, they would have got short shrift.

Needham's help regarding my permission to stay in the UK proved as effective as it was genuinely friendly, which was typical of a devout Anglican, of great integrity and trust. His memory at Cambridge is celebrated in an Institute of Chinese Studies, which he endowed with his archives and where his work is continued by future generations.

RUMOURS ROUND THE 'BACKS'

After nearly one year of painstakingly gaining these concessions it was time for Lord Goodman to wrap up the proceedings, as the most important of all team players.

On 5 April 1973 my five-year-old Romanian passport would have lapsed with no chance of it being renewed by the Romanian authorities in Bucharest. By this time I would have resided nearly five calendar years in the United Kingdom.

William Deer, Professor of Mineralogy, Master of Trinity Hall and Vice-Chancellor of Cambridge University, wrote to the Home Secretary Robert Carr in February 1973 saying that he 'endorses the request of the College for Mr Roman to be allowed to stay in the UK'.

The previous Vice-Chancellor, the Reverend Professor W O Chadwick, Master of Selwyn College, had also made representations in this respect to the Home Secretary, the previous year. Reverend Chadwick was eminently familiar with the abuses of the Ceauşescu régime, as his brother was Ambassador to Bucharest.

The Home Secretary was under pressure from other quarters too. This was a two-pronged campaign, in parallel with the pressure borne directly by David Lane, the Under-Secretary of State for Immigration, who answered to the Home Secretary. There were no great in roads made in these two directions but the pressure was constantly present and ever increasing.

A fresh session with Lord Goodman, in his London offices, gave me the opportunity to appraise him of the latest technical difficulties encountered in making the case to the Home Office. As a result of this emergency meeting, only weeks before my travel documents expired, Goodman aimed his famous salvo in the direction of Lane alluding to 'a minor stir on the Cambridge Backs' and urging the Under-Secretary to consider 'a rethink' in his own department.

The opening of the letter is typically Goodmanesque as he quips:

"I promised Lord Colville, some while back, not to bother your Department for one hundred years, but counting each day of the gas strike as ten years, I can, I think, find some auspicious excuse for coming back on a matter that has been especially referred to me by Sir Duncan Wilson, Master of Corpus Christi College, Cambridge. You will probably know all about it and are fed up with it. It is the case of Mr Constantin Roman, a Romanian subject of Peterhouse."

Lane and his Department must have been fed up to the teeth by this time, but the fear of the reception I might get on my return to Ceauşescu's paradise did not allow me the luxury of slackening the pressure. Goodman goes on praising my potential contribution by stating:

"He is a man of impeccable character and he is clearly determined to belong here and make a significant contribution to our national life."

Then, having made a case for putting me in a special category, Goodman went on by saying:

"What impressed me about this young man, was his absolute obduracy, reflecting an attitude of mind which has clearly developed from strong moral factors."

How did Goodman guess my DNA signature, the inheritance of many generations of uncompromising misers? The inheritance of high moral ground but lost causes? Did he know what holocaust we went through? He did not, but he had that ineffable intuition, which set Arnold Goodman apart from his contemporaries and made him into a formidable negotiator, as he added:

"I will willingly and happily come and talk to you about this, since it seems to me deserving of at least my time, and you, alas, can only be a victim in these circumstances."

FROM ROMAN TO MACROMAN

In the middle of this campaign, whose cross-fire made for a fascinating display of fireworks, I was never quite clear how effective the gunners were on the battle field or which side would blink first. One of the two pre-requisites spelled out in Robert Carr's letter to Needham indicated that I had to have a job offer, so I redoubled my efforts in this direction.

As a British or an Anglo-Dutch Oil company would not touch me with a barge pole, what about a French company? I spoke French fluently and felt every bit a Francophile. French, after all, was my first foreign language proper, before English. Let us try 'Total' Compagnie Française des Petroles!

It was in the days when oil was starting to be discovered in the North Sea and the Scottish nationalists were making early representations to the effect that 'the oil was going to be all ours' (i.e. theirs).

I was seen by Total's Personnel Manger. We conversed in French and he confided that they were going to move their offices to Aberdeen in order to 'placate' the Scottish nationalist feelings, 'just in case' and that the company's new policy was 'to recruit only Scotsmen'.

I was very disappointed indeed by this positive discrimination, but at least I knew straight away that I stood no chance. I tried gleefully to suggest that I was 'ready to change my name, by deed poll, from Roman to MacRoman', but the French sense of humour was not one to laugh at such things.

GERMAN AND CANADIAN OFFERS

After this fresh setback I started in earnest to look for jobs abroad, just in case I could not get work in England: Canada and Germany seemed to be obvious targets. I wrote to Professor Hans Berckhemmer, at Goethe University, in Frankfurt: he told me that his assistant professor had just got the Chair of Geophysics at the University of Kiel and that he had ample research grants for staff. I soon got in touch with Kiel University and by return got a job offer, as Assistant Lecturer, conditional on obtaining my degree at Cambridge. Berckhemmer also suggested that I should apply for a Humboldt Fellowship at Frankfurt.

I had also renewed contacts with my old friend Professor Tuzo Wilson, at Toronto University, who in 1969 had offered me a PhD scholarship in Geophysics.

Unlike Germany, in Canada there was no language barrier and the attitude in the University was less contrite. Tuzo offered me a Lectureship straight away. I soon regained my flagging self-confidence.

FINDING WORK IN ENGLAND

The pointer on the employment barometer started to swing to fair weather, but not yet to my obtaining a job in Britain, which the Home Secretary required to grant me permission of abode. I confided in Roxana, who closely followed my trials and tribulations and who sometimes helped with typing my application letters:

"What is to be done? I have tried very hard to find a job here and no one would offer anything. I might be compelled to accept a job offer either with Tuzo Wilson in Canada, or in Germany where I am offered a Humboldt Fellowship by Hans Berckhemmer, at Frankfurt University."

The idea of emigrating was not espoused with a great deal of enthusiasm by Roxana, as she said demurely:

"You have simply got to try harder, Constantin, and you'll find something in the end."

"It is not for lack of trying that I have got no offer to date in Britain! Don't you remember that I got as far as being advised by BP to 'settle for the fact that the BP Group, as a whole, has no opening for me'? When I read this advice I felt as if the sky collapsed on me, that I was unemployable!"

"Never mind about all this, stop whinging and keep on trying!"

"Simple as pie, isn't it?"

"How come that I haven't thought of this before?"

DAILY TELEGRAPH'S RESCUE PLAN

I was definitely in a 'jam' about the prospects of a job offer in Britain. I mulled endlessly about the handicaps I had: 32 years behind me, with no previous employment history, lack of industry experience, 'excessive' qualifications (eleven years in academia with a PhD), lack of work permit, no permission to stay in Britain and a dubious passport/nationality from a Communist country. This would be enough to damn any applicant forever. Even the unthinkable of 'pulling strings' with the Chairman of BP did not work, or applying for jobs outside geology (Jim Slater). This was a real impasse, with little, if any, prospect of getting out of this intractable morass.

I told David Floyd from the *Daily Telegraph*, about my trials and tribulations. I said that I had made over 160 applications and nothing, absolutely nothing came of it, other than be told that I was 'overqualified'.

David Floyd was Communist Affairs correspondent for Eastern Europe. Prior to his career as Correspondent, David was a diplomat at the British embassies in Moscow, Prague and Belgrade. In Belgrade he was a contemporary of Duncan Wilson and Durrell, and I assumed he figured in the latter's book *Esprit de Corps,* under an assumed name. There was no need to give him further graphic details of the *Communist paradise*, which he knew first hand, and understood my predicament. David Floyd was the author of several political books one of which was on *Romania—Russia's dissident Ally*. He knew both systems perfectly well and he recognized instantly the vicious circle that entangled me, which he proceeded to break. It was the same old chess game—to be played by the rules. David understood the scenario: in order to be allowed to stay, I had to have a job. In order to have a job, I had to have a work permit and in order to have a work permit, I should have had a job offer, which was not forthcoming, because I had no work permit, which....

"Well", he said, "all you need is to have a job offer. Simple, isn't it?"

"Yes, but how? Nobody would make an offer!"

"I will."

"How can you? You are a journalist and I am a seismologist. How could you offer me a job?"

David Floyd, Communist Affairs Correspondent of the Daily Telegraph (seen in profile on the left) and Nobel laureate Alexander Solzhenytsin, recently released from Russia. Photograph by kind permission of the Trustees of the late David Floyd Estate. David, a graduate of Trinity College Cambridge, was eminently qualified as a diplomat and journalist to understand both East and West and as such was a loyal friend and staunch supporter of the Roman cause célèbre.

"Very simple—you will help me update my book on Romania, by doing some research for me."

He made an appointment at the Home Office, took me along, told them about his job offer and asked for a work permit. He told the Home Office official that I 'was best qualified, speaking Romanian'. After a lapse of time I got my work permit.

I told Teddy the news of my job.

"Splendid, what job is it?"

"Doing research for the *Daily Telegraph*."

"Excellent! Excellent! Just as Rutherford told me in the 1930's, 'take any job!'"

The old boy was relieved—history was repeating itself, times were hard, and now he could see me stand on my own feet.

WORK PERMIT

No sooner had the authorities given me the work permit than that precious detail was included in my CV. This proved essential in persuading work-permit shy employers to go ahead and make a real offer, based on my professional potential.

This miracle gave fresh impetus to my campaign to find a job in Britain. I felt I was justified in trying harder now, as it might not be in vain this time. With OPEC getting off the ground and manipulating the price of crude oil, there was an urgent need in the Western world to find more oil in the 'right geographical areas', providing the West with a sure source of energy, without holding it for ransom with exorbitant prices.

The Americans and the British started to look in earnest for oil in the North Sea. The job opportunities for geologists started to improve all of a sudden and with a work permit in my pocket, things were, for the first time, looking rather good.

INDEFINITE STAY

Eventually, the long expected answer came and filtered through to all my supporters in Cambridge and London and I was given leave for an indefinite stay, on the basis of the job offer made by David Floyd, of the *Daily Telegraph*. At a stroke I had a work permit and permission to stay, but not without waiting a further gruelling five months for my papers to be endorsed.

Lord Goodman wrote in his unmistakable style:

"The Home Office have behaved in a most helpful and civilized manner in this matter, reflecting the greatest credit on it."

It was typical of Goodman to twist the opponent's arm and then to make him feel good, to give him credit for yielding to his demands. This was an absolute gem, as it caused many dozens of respectable people to spend time pleading with the authorities in my favour. But there was no time for looking back. It was time to count my blessings and feel free to resolve my other major tasks: finishing the dissertation, obtaining a proper long term job in my own field, which in turn would make me free to sort out my personal life, away from Cambridge.

'I'M AN OKY FROM WISKOGY'

My endeavours to find a job eventually bore fruit, through fortuitous political and economic conjecture. During the early 1970s the Arab oil-producing countries had become more demanding, requiring a greater and greater share of the profits made by the Western oil companies.

Teddy told me: "It was inevitable—they were bound to ask more, after all, these are their riches and they want to profit from them."

The cost of a barrel of oil, to which I was quite oblivious as a pointer to my own employment fortunes, started to rise sharply in a spiral which was to be known as 'the first oil crisis'. This was a misnomer, as it was neither the 'first', nor the 'last' oil crisis: the oil industry, since its early beginnings in the 'Wild West', was periodically bedevilled by ups and downs. As a result of these latest developments, the American and British oil companies started to look in earnest for oil closer to home and in new geological basins, in order to make more hydrocarbon discoveries and therefore be less dependent on the Middle East. Oil exploration expanded and as a result geologists and geophysicists were again in demand. North Sea exploration was all the rage and it was moving from the shallower waters of the southern North Sea (the gas producing area), to the deeper technology of the oil-producing Viking Graben of the northern North Sea. Jack-up platforms, standing on the bottom of the Sea, had to make way for floating platforms, in deeper waters. The technology was changing rapidly.

I read all the advertisements in the *Sunday Times* and suddenly I came across the name of an oil company I had never heard of before: CONOCO. Nearly all American oil companies had the habit of ending their name in 'oco', an acronym for oil company. Well, but 'Con', what could 'Con' be? It was 'Continental Oil Company', with offices in Park Street, which first offered me a job. The company had exploration licences in the North Sea and the midwest of the United States, in Oklahoma.

"Where is Oklahoma?"

I had to look on the map. I was so excited about this offer that I did not even negotiate about my salary. There was nothing to negotiate—I was to be taken on, as a trainee geophysicist and the salary was fixed at £2700 per annum. I forgot to tell them that I had just got married and I had not yet gone on a honeymoon. I had no money for the honeymoon. Besides I was afraid that, if I did not take the job immediately, I might lose it.

When I finally got my first job offer in London, I was absolutely elated to have reached the end of the tunnel, after two years of solid job search: I did not feel bitter—I was too young for such feelings and considered myself lucky to have got so far.

This achievement did not stop my English wife from pointing out the obvious, at the time:

"It took an American company to offer you a job!"

"Well, oil is quite international and Americans have had dealings with it far longer than the British".

"Which Roman Emperor said, 'money did not stink'?"

We said goodbye to Peterhouse. I was elated—I had a degree from Cambridge, a wife, a flat in Chelsea and a job with good prospects. The world was all mine again.

VIVA

The text of the dissertation was growing, but coming to a close.

"Finish it, finish it", Molly would urge me. "You must finish it", Teddy said. I knew that I had to, but what they did not know was that I had more pages of correspondence with the Home Office over my status in the UK, than in my PhD dissertation.

Here I was, finally! I had finished the 'Thing'!

Roxana, who was a secretary in London, having taken, amongst other qualifications, the Pitman course, offered to type my manuscript. I soon discovered that it was not the best idea. Molly found me a cheap and efficient typist near Parker's Piece, who indirectly saved the matrimonial peace. The text was soon bound, but it cost an arm and a leg, which I could ill afford. Molly did advise putting in a pocket at the end of my thesis, containing reprints of my publications on geophysics in English, French and Romanian, of which I a had a few, probably more than my contemporaries. My internal examiner was Drum Matthews and the external examiner was Hal Thirlaway, from the UKAEA in Aldermaston. The viva was to take place in Thirloway's office in Aldermaston.

Drum gave me a lift in his car. The cross-country road from Cambridge to Aldermaston gave Drum the opportunity to acquaint himself with my other activities in Cambridge. He mentioned the translation of the Romanian poem I had put at the beginning of my thesis and which I had published in *'Encounter'*, in the previous year:

I am writing on Earthquake
If some of my words
Slide too far on,
It's the Crust of the Earth that's to blame,
With its lack of stability.

Encounter paid eight pounds for four poem translations, which helped bind my dissertation. However the costs of producing the thesis caused me to have to take an interest-free loan of £500 from a special fund run by the University of Cambridge, which I reimbursed, in monthly instalments, over a period of three years. I was quite grateful for this source of sustenance and I remember that the University required the backing of a guarantor, in order to release the funds. Roger Lovatt, my Tutor in Peterhouse, asked whom I proposed as a guarantor.

"You, of course."

"No, I can't", he said shyly. "You must put somebody who knows you well."

"But you know me very well."

"No, this is no good, it will have to be somebody else than your Tutor."

"But I know nobody well enough to be my guarantor. I presume I will have to forego this loan."

Poor Roger, in the end he explained to the University that I 'knew nobody' and eventually the loan was granted.

'GEOPHYSIST?' NO, 'GEO-PHY-SI-CIST!'

My Romanian passport was due to expire after five years, on 5 April 1974. It had become gradually clear to me that I was driving down a one-way

145

street. There was no 'U' turn, no going back. I would have been harassed, compromised, misused. The few flash points which I had from Romania, whether the conversation with Lipatti in Paris, with the embassy official in Cambridge, with my former professor in Luxembourg, all of these pointed in the same direction—on the 'road of no return'.

I no longer needed an excuse for not wishing to return to a Communist country. Now that I was engaged to be married, my English fiancée, in the best tradition of equality of the sexes, had the right to choose where she wanted to live: not surprisingly, she wanted to live in England. I might have considered settling in another country if I had a job offer elsewhere, but eventually an American Oil Company in London made the move. It was important to foresee what might have happened once my Romanian passport expired and make provisions for it.

My student status in Cambridge meant that I was going to be supported by my College, together with many other Cambridge colleges, by individual dons and by the University administration. The process was not straightforward and it took at least two years of patiently hammering at this issue with the Home Office authorities. The silver lining of this unexpectedly long and tortuous action was that it took a year longer to complete my thesis and consequently I was still at Cambridge when my Romanian passport expired. It was clear that I needed travel documents, other than a passport from a Communist country, whilst I was still a college graduate.

In 1973 I had been granted permission to stay indefinitely in Britain, which was a major breakthrough. I had also obtained a work permit. In the meantime I was advised to apply for travel documents under the Geneva Convention. These were soon granted and I could see already that it was a great deal easier to travel on stateless papers than on a passport from a Communist country. When I went to collect my new papers at the Home Office, I was bemused to see the girl at the desk hesitating:

"Is it geophysist?"

"No, geophy-*si-cist*", I spelled it slowly for her

By now, I was resigned to this kind of spelling confusion and mispronunciation:

"Fancy, the clerk who handles *my* travel documents does not even know how to speak English properly. We will soon be controlled by faceless, semi-educated officials!"

The discrimination had nothing to do with race, but with education. One is always afraid of what one does not know, of what one does not understand.

Thankfully, Cambridge saw that one of their more educated alumni would be granted the right to exist, even though he was, for the time being, travelling on some sort of 'funny' passport.

UNCERTAIN NATIONALITY

In October 1973 I had got married in the registry office in Cambridge. By February 1974 the University Praelector informed me that I was approved for the PhD. The same month I signed a contract of employment with the Continental Oil Company in Park Street. The question of travel documents still remained to be solved. My Romanian passport was now out of date and I had still to apply for British Citizenship.

Under the Geneva Convention I was given stateless travel papers of a 'citizen of uncertain nationality'. The word 'uncertain' had nothing to do with any renunciation on my part of Romanian nationality: on this particular point I was quite emphatic from the outset, when the Romanian Embassy had demanded that I should renounce my nationality and I refused. Clearly, being a Romanian citizen, living in the West had no advantage and only inconvenience—but this was a right of birth, which nobody could take away from me. The word 'uncertain' had to do with the old Communist practice that a citizen, such as myself, who had 'fled', or failed to return, had his nationality taken away from him. Solzhenytsin was the most famous example, but there were plenty more like him, from every Communist country.

What was more extraordinary than the description in the official paper, was the actual format of the document, a kind of A4 piece of paper, folded in four, like some office memo, with a photograph and the Home Office stamp allowing me re-entry to the UK and residence status. This scrap of indifferent manuscript did not appear to command much respect at the border of any country. Quite the contrary, it created panic and puzzle. It might be read upside down and back to front, kept hanging in the air by two fingers, as if it came from some pestilential hospice. It was humiliating, but I decided to look the official straight in the eye, waiting to see his reaction.

In the end, what mattered most, was that I was much freer to come and go throughout Western Europe than I ever was before, on a passport issued by the Socialist Republic of Romania. This was in 1974, some fifteen years before Ceauşescu was shouted down by the angry crowds of Bucharest and subsequently put down in a palace coup d'état. I felt, travelling on these fancy papers, that I had become a 'Citizen of the World' and that eventually I would obtain the due credit and dignity of every self-respecting citizen of the free world and travel on a normal passport, once my roots had found a favourable soil.

SQUARE HAT

I went to my Degree Ceremony in Cambridge during the spring of 1974, months after I had submitted my dissertation and been informed that I was granted the degree. The ceremony was purely symbolic and I was not obliged to go through with it, but how could I forego such pageantry?

147

After all my trials and tribulations I certainly deserved a jolly, as well as a little celebration, with a few good friends from Cambridge and elsewhere. Molly, my indefatigable supporter, was present too. Each student receiving his degree was allowed only one guest in the Senate House, so apart from my wife, my friends congregated outside the main entrance. All students were received in order of precedence, dictated by the age of each college. Peterhouse being the oldest of Cambridge colleges, we were received in first. Dr Hinton ushered us all in, having led us along Trumpington Street, all the way to Senate House. We were about a dozen graduates of Peterhouse, proceeding proudly in our gowns, with the PhD students walking ahead of those who were going to get a BA. As I never bought a hat I had to hire one for the occasion, the typical 'square'. These were not much in use anymore, except for special ceremonies. Dr R Hinton, a Fellow of History at Peterhouse, presented us to the Vice-Chancellor, speaking in Latin. We had to kneel and were pronounced, also in Latin, to have received the degree of Doctor of Philosophy of the University of Cambridge. The ceremony was simple. No cameras or tape recorders were allowed inside. We returned to our hotel quarters, passing through the grounds of King's, across the bridge and along the Backs, which were resplendent with spring colours.

We hired a reception suite at the Garden House Hotel, the same place which was set on fire, some three years previously, by 'revolutionary' students from Peterhouse.

Quite a few continental friends travelled across the Channel for the ceremony. Only the Roman parents from Bucharest were absent: they were denied permission, by Ceauşescu, to travel to the UK, but we were determined not to let these absurd vexations cast a shadow on a joyful event— my parents were with us in spirit and they celebrated the same day in Bucharest. From amongst the supporters at the Department of Geophysics came Drum Matthews and Molly Wisdom, but the guest of honour was Teddy, the friendly, unceremonious, good old Teddy!—I felt that I had learned more than geophysics from him, I had learned to be a Man!

I was his last student, as he was going to retire from Cambridge. Although his health was failing, he chose to take an active retirement, which was made possible by the offer of an Emeritus Professorship at La Jolla, in California, which he loved so much. I owed him a debt of gratitude and I was going to miss him immensely.

CHAPTER 6

LOTUS-EATER

"J'aime mieux être un homme à paradoxes,
qu'un homme à préjugés"
(Jean-Jacques Rousseau
Emile, ou l'éducation)

LOTUS-EATER

In my quest for work and for the right to exist as a free man, I had more than my fair share of difficulties, some self-imposed, others quite artificial and arbitrary, created by the prejudices of blinkered individuals, or bureaucracies. Coming to grips with these handicaps and succeeding in adverse circumstances took up a lot of stamina to the detriment of more creative activities.

I had to face with stoicism the negative effects which the sterile battles with the bureaucratic dragon had on my general well being, indeed on my very sanity. At times, I felt threatened and insecure. I had nevertheless to dismiss such feelings as a mere 'luxury which I could not afford', simply because I had no other option in sight! What appeared quite clearly, in my search for stability, was the need to compensate, to find a reservoir of peace and inspiration, from which I could draw my strength and recoup my wasted energy. I had no need to look very far afield for it, as Cambridge offered a loving and harmonious atmosphere, where the spirit could flourish, the curiosity could be satisfied and where the heart could sing. My falling in love with Cambridge was not a mere lotus-eating experiment, it was more than skin-deep. Cambridge was almost like a mythical mistress, whose eroticism would excite my resolve against the sheer obstacles put in the way by sundry bureaucratic tormentors and moral dwarfs.

Whenever I think of my University town, the image that I instantly conjure up is that of the Wren Library, in Trinity College, seen from the lawns of St John's College, with the lazy punts gliding past. It has to be a sunny day, with clear blue sky and masses of daffodils along the Backs, like an unreal theatre setting. I used to lie on the river bank, which sloped gently into the water, close my eyes and wish that time would freeze in this blissful moment of immense elation. I had to try hard to come back to reality and reassure myself that this was not a dream. I tore myself away from this idyllic corner, to delight my eyes with the gothic silhouette of the Bridge of Sighs, against the Dutch gables of the Library of St John's. Then, following the path upstream, go past Clare College Gardens, to discover further along the gothic perpendicular Chapel of Henry VI. Its construction was, thankfully, interrupted by the Wars of the Roses, which allowed it to be finished under Henry VIII, by which time the splendid fan vaulting was added, the most perfect example of its kind in the whole country. Back into King's Parade, into the secret passage of Queen's Lane, I discovered the medieval charm of Queens' College, another royal foundation. Across the river, along the Mathematical Bridge, back into Silver Street, I would watch the effervescent weir full of busy ducks and the starting point of many a punt trip up and down the river.

Soon it became clear that the Universe was a little wider and that there were other colleges along Sidney Street and St Andrews Street and beyond. At this point, the College properties became more interwoven with the town proper, until Academia became more 'diluted' and the Town took over, beyond Magdalene and Jesus College to the North, Peterhouse and Downing College to the South. The newest Colleges were more 'isolated', in other words completely surrounded by non-collegiate neighbours, such as Fitzwilliam and New Hall in Huntington Road. Here one had to struggle up hill, on a bicycle, past Castle Hill. The College buildings were modern and had a campus-like atmosphere, quite different from the more spiritual, even quixotic feeling of the older colleges in town.

Market Hill, near Great St Mary's, represented a 'kernel' of Town which had survived so near to the heart of College-land. On the whole, we knew that the Town did exist, but we had no real need to have much to do with it. There were the odd visits to the market, the few shops in Petty Curry, Green Street, Sidney Street and St Andrew's St. Even there the book-sellers, the tailors, the stationers and other small traders would be geared up for students needs and appeared to be mere appendages of academia. 'Fitzbillies', the confectioners, thrived solely on the College clientele. So did Edward Leigh, the photographer in King's Parade, David the anti-quarian bookseller, the Greek restaurant in Rose Crescent, 'Lunn Poly', the travel agent in King's Parade, and others. By the 20th century University restrictions had relaxed and pubs could be frequented by undergraduates, hence the Little Rose, opposite Peterhouse, The Eagle, Hobson's old coach inn in Bene't St and the Blue Boar in Trinity St were nearly all dependent on students. Part of Watson's gestation of the DNA theory had to do with The Eagle, which was only yards from the Cavendish. In the summer, when the undergraduates would disperse, they would be replaced by tourists.

There was an inevitable love–hate relationship between Town and Gown, though it was recognized, grudgingly, that they both needed each other. There were the occasional punch ups, when the odd undergraduate would be beaten up by the lads in town. But nothing as acute as the riots of earlier centuries, when Colleges were nearly burnt down by the mob, protesting at the demolition of their houses to make room for royal foundations, such as King's, when half of the heart of the city was torn apart.

BISHOP'S MONEY
Peterhouse has the oldest Hall in Cambridge, going back to its foundation in 1284. The Hall was restored in the 19th century, when it was decorated by William Morris. It could take up to over one hundred undergraduates, but as their number grew, two sittings were introduced and eventually a self-service system. Formal dinners got fewer and attendance was no longer compulsory. However, as meals were heavily subsidized from college funds and benefactions, it made sense for all students to eat in College. There were, in all, some 240 undergraduate students at Peterhouse and 60 postgraduates.

The College had charitable status, being a seat of learning, which gave it considerable tax advantages. This meant that all donations or 'benefactions' were exempt of tax and so the wealth of the College could accrue accordingly. Revenue was not taxed either, whether it came from arable land let to tenant farmers, or from rent of property in London and elsewhere, or indeed from financial investments in the City. The Bursar, who was accountable to the Master and Fellows, the custodians of these 'livings', looked after the affairs of the College. Peterhouse, not being a Royal foundation and being one of the smallest of Cambridge colleges, was not the richest either. But its benefactions had accumulated in value over seven hundred years of its existence. Like other Colleges in Cambridge Peterhouse had certain benefactions intended solely for scholarships, with specific clauses attached to them. The Research Scholarship, which I together with three or four other postgraduates was granted by Peterhouse, had its source in a fund established by the Bishop of Ely in the 13th century. It therefore gave me tremendous pleasure to acknowledge this, especially to those who would say:

"This is the British Taxpayer's money".

"No", I would retort, not at all, this is Hugh de Balsham's money".

"Who is he?"

"Why? The Bishop of Ely, of course! He lived in the 13th century, when your ancestors probably were herding the swine".

HOSTEL ROOMS
Sir Nikolaus Pevsner, the distinguished German art historian who came to England before the war, did not have many kind words for the hostel of Peterhouse. When he came to describing it, in his book on the 'Buildings

Peterhouse, the oldest Cambridge College, founded in 1284, offered a four-year Research Scholarship which allowed me to obtain a PhD in geophysics. The college is seen from the author's windows from across Trumpington Street. (Photo by Constantin Roman 1970.)

of England, Cambridgeshire', Sir Nikolaus referred to it as 'an inoffensive neo-Georgian style and might be a post office'.

Being a postgraduate Scholar I was given rooms in Sir Nikolaus' 'Post Office', otherwise known as 'The Hostel'. This was a 1920s building with a red brick elevation, Georgian style sash windows with blue shutters, for a dozen students. My suite contained a small lobby, leading to a comfortable sitting room-cum-study, with a fireplace and windows overlooking the street. I had a complete wall to accommodate my tectonic map of Eurasia, which came in 12 sheets. Off the sitting room, a door led to a small narrow bedroom, with enough space for a single bed and side table. Opposite my rooms, across the corridor, was the kitchenette, with a single gas ring on which I would cook three-course meals when I entertained guests. The fireplace was the classical setting where I would take pictures of my female visitors, who were not allowed to stay the night.

The hostel had the advantage of offering a satisfactory degree of independence, compared to the régime of the undergraduates living in College, across the road, who in 1969 still had to be in by midnight. Eventually the rules relaxed a little and one could enter the College through the only open gate, by the porter's lodge, between 12 and 2 am. In the hostel we had our own front door key, so we were not required to keep these hours. But the Porter, across the road, kept a vigilant eye on us and so did the 'Bedders', who were the two old women who came in every morning to make the bed and clean the rooms. They would report anything suspicious to the Porter, who would filter the news to the Senior Tutor. Very little would go unmissed and if I had no feedback it was usually because postgraduate status had more lax, unwritten rules than the undergraduate strictures. The advantage of living in the hostel was its proximity to College. A few leaps across Trumpington Street and one would find oneself having dinner in Hall. From my windows I could see the College's 17th century Chapel, with its beautiful stained glass, after Rubens, as well as the 16th century corbel window of the Perne Library.

'CROQUETING'

Peterhouse had extensive grounds between Trumpington Street and the river Granta, stretching all the way to the Engineering Department. The Fitzwilliam Museum was built in the 19th century on land given by the College, whose gardens backed onto the Museum. On the other side of the garden a massive medieval nine-foot wall, parallel to Trumpington Street, separated Peterhouse from the common land along the Granta. In the old days, this wall was intended to stop the students climbing over into the College, once the gates were locked at night. The University 'Bulldogs' did the rest of the job should an unfortunate student remain locked out and unable to scale the wall. To make things more discouraging the wall had embedded on its top broken glass from bottles of different colours, not a very comfortable place to grab if one had to negotiate a passage and climb over it. At the time of my stay in Cambridge, a popular title of a book on sale in town was *Night climbing in Cambridge*, giving tips to those undergraduates willing to risk the integrity of their trousers, by grating them on the broken glass of high perimeter walls.

Squash became one of my favourite sports, as it involved the expenditure of the maximum of energy in a minimum of time. I was playing it three times a week, usually finishing late at night, just before the College bar in the JCR closed and I could order a pint of beer. The courts were behind the newly erected William Stone building. This was a recent tower block, with splendid views over the river and beyond. It was built by Sir Leslie Martin and it was perhaps the most successful piece of modern architecture at Peterhouse. The construction was named after an old boy who died a nonagenarian bachelor and was one of the most recent and most generous benefactors of the College. His ashes were buried in the

College Chapel. The fact that William Stone left money to his old College was the result of a diligent and persistent public relations exercise by the College, which despatched to London, every week, one of the fellows 'to keep in touch'. In the end their efforts had borne fruit.

The Scholars' garden had a beautiful shrubbery and herbaceous border and an extensive croquet lawn. Here I was taught the rules of the game, on a beautiful summer afternoon, with a group of graduate students. My difficulty was that I could not hit the ball straight with the mallet, as I was shaking with laughter at what I thought was a typical English scene from a Miss Marple story, by Agatha Christie.

PUNTING ON THE RIVER

The river is so lazy that one could hardly guess the direction of its flow, punctuated by the gliding of the swans, or the more boisterous mallards of the Millpond, intermingling with the elegant, almost aerodynamic punts. These are commanded and steered from the back by means of a ten-foot-long wooden pole with a metal prong at one end, which helps to push the punt forward. As the punt advances, the pole has to be swiftly and skilfully plucked out of water, hoiked into the air only to stab again the river bed, to repeat the movement a few yards further on. Occasionally, as the river bed is thick with mud, the pole gets stuck as the puntsman tries to lift it. Then, in a split second, the punt glides away, whilst the unlucky driver clings on to the pole with an uncertain balance and eventually plunges, like a pendulum, into the water.

The Fellows' gardens or the Master's gardens intersperse the college buildings on the river, and are linked across the Cam by elegant stone bridges to more gardens bordering the 'Backs'.

The Backs retain, however, the feeling of a more natural woodland aspect, which in Spring is rich with the colour of crocuses, daffodils and narcissi. To sit on the left bank of the river, amongst the daffodils, contemplating the ancient buildings on the right bank glistening in the afternoon sun, is an utter and most complete delight. This was not the vulgar picture taken from a chocolate box, but a blissful piece of paradise, a corner of the Universe where the clock has stopped, yet, at the same time, it gives reassurance through its continuity. Kruschev, once a visitor in Cambridge, broke his speechlessness by making an enquiry closest to the only subject he could claim knowledge of, which was agriculture. So, he asked his host, the Provost of King's College:

"What is the secret of this perfect lawn?"

"Ah", said the Provost, "You see, we have a seven hundred years old culture."

The story may be apocryphal, like so many anecdotes that punctuate Cambridge history, yet it befits this august place of learning.

Later on in life, when I bought a home, I would first choose a lawn with perfect herbaceous borders and the house as a subsidiary: something to look out from, rather than to look in. The long contemplative moments on the river Cam have a lot to do with these latter choices.

COLLEGE GARDENS

The rivers Granta and Cam bring the fens into the 'Backs' and the Backs bring the countryside into the College gardens, almost like a natural succession, only more orderly and manicured.

Clare College gardens are better seen and enjoyed from the river, others are a little bit more 'secret', like the Fellows' gardens at Trinity Hall, or the Master's gardens of Queens. There are, of course, many completely enclosed gardens, the existence of which one can only conjecture about.

Each College, whether on the river or not, has a series of gardens, each trying to outdo the other: each of them having its individuality and charm. The element of 'surprise' leads one from one enclosed garden to another, through a gothic gate, or a topiary archway. The pond gardens of Emmanuel, the rose garden of Downing, the wonderful ironwork decorating the garden gates of Trinity and St John's, the Henry Moore bronzes on the lawns of Churchill College, they are all refined examples of English landscape architecture, which add to the distinction of College buildings, of which they play an integral part. Nothing stops abruptly, just flows, continues into the open, which makes it difficult to separate one from the other.

Yet there are gardens further afield from the main College buildings, tucked away secretly, like the Fellows' gardens at Trinity, across the Backs near the University Library. One would enter the huge expanse of the Old Court, through the screens, past the Wren's Library, across the river and along the alleyway bordered by majestic elms, through the elaborate iron gate at the back, crossing the road to reach a smaller gate, on the other side. This would have been open only for the Fellows of the College, but hardly anybody ever came here. I would sneak in, like a thief, as I knew that I had no right to be there, yet the temptation of enjoying complete peace, away from the crowds, was too great to resist

Occasionally, I would see the furtive silhouette of Lord Adrian, the Chancellor of the University, crossing the garden. The archetype of the Scientist, his demeanour was not aggressive and always displayed immense humility.

"Good afternoon, Sir", I would say if he passed too closely to the bench where I was seated. He would nod courteously and would answer "Good afternoon", although we never met formally: I felt guilty at interrupting the great man's train of thought, doubtless dreaming up some novel application of bio-currents.

By contrast to countries behind the Iron Curtain, Cambridge was an epitome of civilization, the perfect environment which gave the world many outstanding scientists. It probably had the highest concentration of Nobel Prize winners per square mile, anywhere in the world.

Cambridge would elevate the spirit, whether by contemplating the gardens, the people, or the buildings. I often raised my eyes to the skies from the bench in the Trinity College Fellows' gardens and felt like singing a song of praise to the Lord, for allowing me to be a small part of this paradise. Alleluia! Alleluia!

For a split second I believed that the gardens of Gethsemane were in Cambridge.

There is, of course, no Cambridge garden without its garden party. Everybody who is anybody and has a worthwhile garden, gives a garden party, at least once a year. Even anybody who is somebody who hardly has any garden at all, like the minutest town garden of the size of a handkerchief, still gives a garden party. Sometimes, even anybody who is nobody, who hardly has a garden at all... Or, would the garden matter, if that somebody was a nobody? No, it wouldn't, because the people would still come. What would matter most would be the garden, the drinks, the cucumber sandwiches and the company, in that order. The host mattered less, provided that he offered the venue, the drinks and the sandwiches. The company would automatically form itself. Yes, sometimes it mattered if the host or hostess were glamorous.

Fresh in Cambridge, my name was not yet on the social circuit, so I did not know all the ins and outs of what made a garden party animal, or in other words, what characteristics one had to have to be invited, to a garden party. However, as 'curiosity killed the cat', I soon got to be invited merely for exhibition purposes, for friends and acquaintances to find out if I was in any way 'related to Dracula', or in what way I might be different from a unicorn.

"Do Romanians have horns?"

"Yes, Ma'am, they have alpine horns."

"?!?!?"

"Oh, sorry Ma'am, you mean some sort of devil-cum-vampire with horns, like a minotaur? Yes, they do, but they don't display them. Romanians can be very shy and retiring, they have retractable horns, like the snails. If you egg them on too much and they get terribly excited, their horns might come out in anger and you can pray for mercy not to be impaled."

During the late spring and summer the gardens of Peterhouse were particularly beautiful, starting with the aconites, the crocuses, the narcissi and daffodils, naturalized in the Deer Park and elsewhere. On the croquet lawn of the Scholars' Garden one would often be invited to a garden party with a buffet lunch. It was a splendid occasion, with the undergraduates wearing the boat club jackets and ties and straw hats. It reminded me of the dated photographs of my grandfather's youth, sporting the same shape hat. Whilst in Romania, those times were long past, in England, thankfully, these traditions were preserved. What could Romania have become had the Russian armies not brought Communism with them? We would have lived better things!

SHEPHERD'S GOWN

Soon after the beginning of the first Michaelmas term, the Senior Tutor informed all first-year Research Scholars to be ready to meet the Master, at his Lodge, at a set time, for drinks. We were asked to wear the gown over a dark suit.

The Master's Lodge was only next door to the hostel, but what an excitement to reach through the ornate iron gate and enter this elegant building!

"What would it look like behind this charming Queen Anne facade, described by Pevsner as 'very plain'?"

This was a tall, double pile building, with a wooden oak staircase leading to the first floor sitting room, where we were received.

The Master was a retiring character, almost to the extreme of shyness. His wife was doing most of the conversation, showing us her collection of netsuke. Her husband operated more quietly, approaching individual scholars to ask a few banal questions, almost as the Queen would do at public functions. He knew that he had to go through the motions of this courtesy call and he had to show that he was interested. In fact the person who appeared to enjoy it most was Gretta Burkill. She proceeded to tell us of her concern for the welfare of the graduate families and that of Visiting Scholars in Cambridge. Her interest was genuine and not confined to pious words, she actually *saw* that things *did* happen, that they *did* come to fruition. She told me how she approached Wolfson (later Lord Wolfson) to part with some cash, in order to build the Graduate Centre, overlooking the river near Peterhouse. She had entertained him in Cambridge and over dinner the question was put to him. He obliged. The building was now there for all graduates to enjoy. It was Mrs Burkill's hard work behind the scenes which did it. As I was ensconced in this musing I hardly realized that our visit was coming to a close. The Master courteously accompanied us to the door. At this point, I put my winter jacket on, which the Master could not help noticing. My jacket was made of thick white woven wool, with long hair, almost like an Afghan coat, which was popular with young people at the time. Mine, however was from Transylvania, woven for a Carpathian shepherd by his wife. It was *not* the sort of attire to bring to the Master's Lodge, but I did it with a purpose and the opportunity finally materialized, as the Master remarked:

"What a beautiful jacket, what is it?"

Was it shock, was it amazement at my daring, or was it a subtle tease?

"Master", I replied, "this is the Gown of the University of Bucharest."

He smiled, rather more broadly than earlier that evening. Now he would remember me well!

BLACK TIE AT MAGDALENE

Another Cambridge ghost from history, rather more revered than Cromwell, was the diarist Samuel Pepys, who bequeathed his important collection of books to Magdalene College, in his will of 1703. The *Biblioteca Pepysiana* could be seen through the archway of the screen of the First Court. This is a remarkable building, with gabled mullioned windows and classical Ionic columns and balustrade, a quaint mixture of styles, so typical of English architecture, where the elements of an older style still lingered on as a new style was being introduced.

The Pepys Library is famous for housing, amongst others, the original manuscripts of Pepys' own diaries, which were written in a coded script, not unusual in the 17th century. It was the task of each Pepys Librarian at Magdalene College, a very prestigious appointment, to produce a new annotated edition of Pepys' Diaries. This was more like a lifelong work.

At the time of my studies in Cambridge, the Pepys librarian was Professor Robert Latham.

One afternoon, in my first term, I got a desperate call from Bob Latham through the Porter's Lodge at Peterhouse: would I 'please call him urgently'. I was out that afternoon and only returned at 6 o'clock. By that time Latham had rung twice. The matter seemed urgent enough for an English gentleman, who hardly knew me, to be so insistent.

"How can I help?"

Would I please "come to a formal dinner in Magdalene, that very evening? It is a black tie affair. Have you got a black tie?" he asked anxiously, thinking that I might wear a Romanian national dress, like King Carol's 'masseurs' at King George V's funeral.

"Sure", I answered, not knowing that a 'black tie' was in fact dinner jacket and bow tie.

In any event, Father's dinner jacket was still in Romania and rather difficult to dispatch by post that very evening. As for a bow tie, I had not acquired one yet and in any case I did not know what a 'black tie' was. I simply imagined it to be an attire for a formal occasion, such as a funeral, where one would wear a black coloured normal long tie.

Far more fun to wear the Romanian flag colours, as I had a tie with blue, yellow and red! Surely, this would do for a smart occasion, such as that of Magdalene!?

With no dinner jacket and no bow tie I must have looked like a Martian in the rarefied surroundings of a very serious affair, which was the formal dinner at Magdalene, when Fellows would bring their guests. Each Fellow would try to outdo another by inviting somebody special, like a High Court judge, a Nobel Prize winner, or a gliteratae, or indeed a well known civil servant, churchman or politician, like a bishop, or a Prime Minister in the making.

The most senior Fellow's guest had the honour of sitting at the High Table at the right of the Master of the College. In this case Latham being the most senior Fellow, I was seated next to the Master. Latham nearly fainted when he saw me in my attire, but he put on a brave face. Obviously his choice of guest, who should have been the summum of wisdom that evening, must have let him down and it was paramount for him to 'fill in' the vacant place, especially that it was next to the Master. He must have tried very hard and had not succeeded in finding a suitable extra and so in desperation called on his Romanian friend to 'stand in'.

"Surely Constantin will not be offended if I called on his services at the last minute?"

"No, not at all."

I was quite relaxed about the occasion, as I was not there to impress anybody. The Master did not appear to mind at all how I was dressed and he was most courteous and polite. Of course, being Romanian DID indeed help the occasion, because I was 'not supposed to know' what the proper manners were, so I was automatically excused for my outlandish outfit!

Magdalene's tradition puts the College, together with Trinity, in a special category of its own. These are the only two Cambridge Colleges where College Fellows do not elect the Master. At Trinity, the Master is a political appointee, chosen by the Queen on the advice of the Prime Minister. Magdalene, which was refounded in the 16th century by the family that owned Audley End, had its Master chosen by whatever family owned that mansion.

This had certain advantages, as the Fellows were resigned to the idea that their Master was always going to be an outsider. At the same time this tradition of an external appointee made life a lot easier, as it deprived the College of the usual turmoil surrounding the machinations and intrigues linked to the choice of a new Master. The election of a College Master was quite a Byzantine affair, not unlike the election of the new Pope by the conclave of Bishops, in this case the Fellows of the College. No such electioneering joy at Magdalene! Here the Master was W Hamilton, a Scotsman, with a blunt sense of humour, all too ready to poke fun at the unsuspecting and terrified College Fellows, most of whom were of a venerable age, or so I thought.

Euthanasia was the buzzword in the national press the day I was asked to dinner in Magdalene. This gave the Master a field day:

"What is your opinion on euthanasia?" the Master would ask pointedly an old college Fellow across the dinner table.

"Do you think that the age of 65 is a reasonable threshold for euthanasia?"

"Well", the Fellow mumbled getting all purple in the face, "I have not given it enough thought."

"Well, you must!" said the Master in a determined voice, "I myself am 60, you must be over 66", then, without waiting for any further confused answer, he chose as a target another poor Fellow.

It was like stalking deer in the Highlands, except that in this particular instance, instead of deer, these were 'old dears'.

It was a memorable affair for me and I am sure for Robert Latham as well. It was the last invitation he extended to me to a black tie occasion.

GIPSY'S DINNER JACKET

The 'black tie' functions to which I was invited in Cambridge were growing in number. I should have had a dinner jacket of my own. Hiring one was expensive. Buying a new one was prohibitive on my grant. Suddenly I remembered that Father still had his dinner jacket in his wardrobe in Bucharest. He had no use for it any longer, as Communism made it a point of principle to be a great leveller. Well, not quite. Some people were more levelled than others: our family was squashed. We were flat on the ground. Dinner jacket functions were abolished in Romania along with the Monarchy in 1947. We had for the best forgotten about dinner jackets, tail coats, top hats and other 'Capitalist attire'. These I could see in family photographs, before the War. But now nobody could wear them, we even had a vested interest in hiding them, as unloved witnesses of pre-revolution times.

"No point in giving proof of our past glories."

I remember, as a child in post-war Communist Romania, being allowed to play football at home with Father's top hat:

"As it had no use—it may just as well offer the child some joy", Mother would venture an excuse for this extraordinary entertainment.

Gipsies in Bucharest called at each home asking for old clothes, which they would barter for kitchen pots and pans. In an impoverished society this was a useful trade. I remember poor Mother tried for ages to swap Father's dinner jacket for some frying pans, but the gipsies would not take it: it had no commercial value.

The point was that the blessed, unloved dinner jacket still pined in Bucharest and I desperately needed it in Cambridge. They were glad to send it over and get rid of it.

Richard Fordham asked me to dinner, with his parents, at the Pitt Club and enquired if I had a dinner jacket. I said yes, to which he remarked reproachfully:

"You bought yourself a dinner jacket?"

"No", I retorted defensively, "This is Father's."

"Do you mean to say you have dinner jackets in Romania?"

GROCER'S DINNER

Maurice Cowling, Fellow of Peterhouse, was the doyen of College historians. Peterhouse abounded in historians, but Maurice was rather special. He had razor-sharp humour and power of analysis. He was a prolific writer, he taught (actually DID teach, as opposed to remaining silent at tutorials) the undergraduates. His rooms in Fen Court were bursting with books and manuscripts. He was also, I presume, a right-wing Tory, which made the College rather polarized, because in those days left-wing sympathizers were known to exist within College precincts.

Maurice had invited Ted Heath, then Prime Minister, to come to dinner in Peterhouse. This was a private function, for the Fellows of College only, with perhaps a few additions from amongst Fellows of other Colleges. The dinner was served in the Combination Room, rather than in Hall, which would have been shared with students in term time. Secrets were hard to keep in Peterhouse and Mr Heath's visit in College was of the open variety. This gave left-wing students the opportunity to organize their protest.

As Ted Heath arrived, banners were unfurled from the attic windows of the Old Court, slogans were shouted, cartoons lampooning Heath as a 'grocer' were displayed, red banners appeared in sight. The kernel of student agitators was not large, but vociferous.

The Combination Room had windows overlooking on one side the garden, where students had no access, but on the other side the windows gave onto the Old Court, where students could not be put out of bounds. These windows offered no insulation to the diners from the excruciating racket outside. However, this havoc was made worse by the blaring of the 'Worker's International' song, through loudspeakers mounted on the roof of the Old Court. The dinner went on, but the conversation inside must have come to a halt. Maurice was shaken by the reaction and probably very embarrassed and angry, but powerless. The most embarrassed of all was Mr Johnson, a Conservative councillor for the City and Kitchen Manager at Peterhouse, who later became Mayor of Cambridge. His faith in 'the Petrean being a Gentleman' was deeply shaken by these events. Ted Heath was unmoved. He must have seen worse. He didn't show his feelings.

In the midst of this entire circus, a solitary figure struck an unfashionable stance in support of Ted Heath. This was a shy undergraduate in History and tutee of Maurice Cowling: his long hair framed a face lit by beady eyes and prominent lips. Against the tide of protest meted out by his fellow Petreans, he needed guts to swim against the tide, qualities one could not easily discern, because of his innocent looks. His name was Michael Portillo, later to become a Cabinet Minister.

IN MEMORIAM BENEFACTORUM

Peterhouse certainly had the oldest array of benefactors, going back to Hugh de Balsham, the 13th century Bishop of Ely. A whole series of buildings, properties in the country and in London, investments in the City and the like provided the College with its tax-free income. All these investments had the same common source—private patronage. Some benefactors gave money for specific purposes, such as building a new Library or Chapel, new lodgings for the Master, or endowing a scholarship, or kitchen fund. It was fit, therefore, to remember this generous largesse in an annual Memorial Service, followed by a College Feast.

This was a splendid occasion, on which the Masters, Fellows, Scholars and Commoners would gather to partake in the service and ceremonial dinner. For some of the senior Fellows this would have been the only occasion they would attend in a year. During the Memorial Service, an interminable list was recited of names of people who had showed their munificence to the College, during its 680 year old history. The names of the Bishop founder and various former masters filed past, including those who sent the College silver to Charles during the Civil War, and others who changed their religion from Catholic to Protestant and back again, during the troubled reigns of Mary and Elizabeth. Puritans and sinners were all dutifully remembered during the service. I wondered what would have been the minimum benefaction to the College in order to be eternally remembered? Would the sum be index-linked to ensure that one was not dropped off the Memorial Service list through depreciation?

After the beautifully sober service, the congregation would leave the 17th century chapel to enter the medieval Hall, where the Feast would take place. The Master presided over the high table, surrounded by past Masters and Senior Fellows, in a strict precedence order. This was the only occasion during the year when some of the undergraduate and graduate Scholars of the College were also invited to seat at the high table. The more junior Fellows and Research Fellows sat with the bulk of the undergraduates, at the other refectory tables, in the medieval Hall. The College silver was polished for the occasion and candlesticks glistened in the wick light. An array of knives and forks, four deep, was placed on either side of each plate and an army of college servants, dressed for the occasion, served sure-footedly.

Had the College Feast ever been accompanied by music for the occasion, during medieval times I wondered.

We had a seating arrangement plan and I was informed in advance that I was to be placed at the High Table, seated opposite the Master. Most importantly of all, on either side of me were seated two of the most distinguished College historians—Sir Herbert Butterfield, a former Master, and Sir Dennis Brogan. Both these worthies were in their late sixties and already retired from University, but not from active life. Dennis Brogan was an expert on American history, a subject about which I knew precious little.

Herbert Butterfield's scholarship was based on medieval Christian history, an even less tangible subject of conversation for a Romanian geophysicist.

Why did they put me there, I wondered? Probably because of the surprise effect that I might have, compared to the more predictable suburban student.

I soon was to discover that Butterfield's personality was very engaging with his mild manner and benevolent smile, which dissolved into a natural humility. He was a student of Temperley, a historian and former Master of Peterhouse, whose interest was East European history. Temperley was a friend of Iorga, the Romanian historian, who became King Carol's Prime Minister and was assassinated by the 'Brown Shirts', in 1940. Iorga was a Scholar of world repute and visited Cambridge as a guest of Temperley at Peterhouse. It is on this occasion that Butterfield had met Iorga.

Turning to my right, to Dennis Brogan, I attempted a safe line of conversation, mentioning the link between the Romanian monarchs and the British Royal family. Brogan came out with the widely believed story, misconceived by the British tabloid press, according to which:

"King Carol of Romania came to the funeral of his cousin King George V accompanied by his masseurs, as he was in need of physical support after a night of dissipation."

I told Brogan that the idea of 'masseurs' came from the fact that representatives of the Romanian political parties, including the Peasant Party, whose members wore a national dress with white tight trousers and long shirt, accompanied Carol.

Fancy how this anecdote still had currency, some thirty years on! Butterfield would have never mentioned this story to me, even if he knew it, simply out of courtesy.

The time came to drink from the ceremonial silver chalice. This was passed round the high table, starting with the Master who would drink *'in memoriam Benefactorum'*, bow to his right, then to his left and pass the cup on to his neighbour, who repeated the gestures and the formula in Latin. Burkill was quite amused to notice my missing the drink, between two ceremonial bows, to Sir Herbert and Sir Denis respectively, as I was quite flustered by the occasion.

YOUR BEATITUDE

Beyond the formal Graduate Society dinners, I had to think constantly of new and more exciting venues to attract as wide a net of members as possible. Some of my contemporaries bemoaned discretely the lack of opportunities for meeting students of the opposite sex. I could not be expected to organize a 'blind date' opportunity for them, but I did write to my counterpart at Girton, asking her if the Peterhouse Graduate Society could come to tea: quite unheard of in both colleges, but this is how new traditions are started.

The President of the Graduate Society in Girton was an American and she probably shared with me the mischief and the quest of the singular. Having put my reputation on the line, I discovered that my colleagues did not believe their luck and shied away. I could not secure a huge crowd, just a few bachelors who were not too embarrassed to be bussed in, as we had to hire a minibus for the occasion.

"Fancy Constantin is bussing us to Girton, what a laugh!"

If our bachelor tea party did not seem to be a hit, it certainly did the rounds of the College. I did not mind a bit the gossip or innuendoes and moved on to the next target.

I had heard much about Peterhouse's relationship with the bishopric of Ely and was intrigued by the fact that the Bishop was called a 'Visitor'. By ancient statute, if the Fellows of the College had an intractable dispute, the Bishop would effectively step in and act as mediator. It was a perfect solution in a conclave which seemed to be rife with disagreements, especially prior to the election of a new Master.

If the 'Visitor' had such College prerogatives to come and "visit" us, what about taking the road in the opposite direction and 'visit' him, for a change?

I thought it might be a wonderful outing for the graduates and their spouses to go on a 'pilgrimage to their roots' in Ely, the source of our foundation. However, as a preamble, we had to be received by the Bishop. I had no hesitation—I would write to the Bishop of Ely and say that we would like to call for tea. But how should one address the Bishop? I presume I could have asked my Tutor, or anybody else in College, but I thought it perfectly all right if I used the same formula intended for the Romanian Orthodox Bishops, so I posted the letter to: 'His Beatitude the Bishop of Ely'.

'His Beatitude' was never addressed like this in his life. He had not even hoped for it in his wildest dreams. How could he decline such a request from an 'Orthodox Brother'?

Bishop Roberts answered in the affirmative.

The visit was on.

It did not make sense to bus in the graduates in the same way as I had done in Girton. This was a serious, more decorous affair. It had to take the aura of a pilgrimage, like that of Santiago de Compostella. Instead of covering the road in half an hour, we would take to the water, very much like how the old Bishops of Ely would have reached Peterhouse from the fens; along the canals, up the rivers Cam and Granta. We hired a boat for the occasion, which would take three hours across the lazy locks that control the waterways to Ely. This would give us ample time to cogitate over more spiritual things, rather than just rush by motor to the ancient cathedral city. 'Spiritual' it had to be and I was determined that it would be pregnant with 'theological meaning'. I took care to order for the occasion a harvest loaf, some six foot in diameter, decorated with the College arms and the date of its foundation. We were going to break the bread, share it with the Bishop of Ely and ask him to bless the red wine, which we would drink with him. Never mind if it was going to be four in the afternoon as we arrived: wine and bread it would be!

The Cathedral silhouette gradually appeared on the horizon, at the beginning quite faint, in shades of pale grey and blue, almost like a water-colour by Turner. It loomed gradually ever larger, until we disembarked on a bank not far from it. Then several of us held on to the round loaf, which must have been quite a sight, as we approached the Bishop's Palace along the streets of Ely.

Bishop Roberts had strong ties with Cambridge. He used to coach one of the boat clubs on the river, which involved the bishop riding a bicycle on the perilous, narrow path along the river Cam, whilst, at the same time, blaring into a portable loudspeaker to encourage the undergraduates to row faster. He would not wear his mitre or purple robes on such occasions.

On our arrival, the Bishop's son, who was a student in Archaeology at Cambridge, directed us on a brief sightseeing tour of the Cathedral. After this we joined his father in the palace gardens. The Bishop greeted us on the lawn, holding his little chihuahua in his arms. It was the first time in the Cathedral's history that Peterhouse students paid a social call on their Visitor. He did not expect such a large attendance and could not help noticing the ceremonial bread decorated with the arms of the College, which we laid on the lawn. The Bishop enquired prudently:

"Why were there two dates 1971 and 1974 on the bread?"

We explained: "As the date of the foundation was not certain as being either 1281 or 1284 we will come again to Ely to pay him a visit in 1974."

Not being able to contain her excitement, Mrs Roberts appeared in the gardens. I was thrilled too, because I did not expect that there would be a 'Mrs Bishop' around. In the Orthodox Church Bishops were celibate and therefore recruited from amongst the monks. More recently, since the Second World War, Orthodox priests who became widowers could also qualify for a bishopric, but Bishop's spouses were inconceivable.

Dorothy Roberts meant well—she produced the Visitor's Book for us to sign. We in turn produced the symbolic bottles of red wine to drink with 'His Beatitude', but before anything so meaningful could happen, some clumsy hands spilled the red wine on the Cathedral's Visitor's Book. Mrs Roberts was not best pleased, *mais ça arrive dans les meilleures familles*, and here we were spilling the 'blood of Christ'. On this memorable note we hastened to a waiting coach, which brought us within half an hour to Cambridge. At least our visit left an indelible mark.

LORD DEWAR'S RESCUE

No sooner had I accepted, with great glee, my Presidency of the Peterhouse Graduate Society than the style of leadership had to change. New blood was needed to inject some tonus into the proceedings and I was determined to encourage more social contacts amongst its members, by creating more diverse and interesting venues. As much as I would have liked to, I could not easily ask the College Housekeeper to change curtains or fitted carpets, or indeed the furniture in the Graduate Common Room. I decided instead, in close consultation with James Thring, a student in architecture, to create a bar area with a kitchenette, hidden behind a wooden screen. A new notice board was installed and the room was extended to the east to include an adjoining space and in the process we discovered an exciting 15th century Gothic archway, hidden in the masonry. This was opened and preserved, enhancing the character of the room.

Still, I felt that the room lacked warmth and could do with some period paintings. Rumour had it that there were some College paintings 'hidden away' and I was determined to inspect and see if any might be available for our enjoyment, to cover the barren walls of the GCR.

I wrote to Professor Clark, an archaeologist who was in charge of the paintings held in storage. I did not expect to find any long-lost Rubens, as I had no illusions that the best paintings the College had were actually hanging either in Hall, in the Master's Lodge, or the Fellows' rooms. Still, it was worth trying.

Graeme Clark took me to the William Stone building and there, in its dry cellars, was piled a series of oil paintings, which I felt were unfashionable, unloved and forgotten. I chose two portraits—one of them was a Victorian oil of the Archbishop of York. Poor fellow, he had collected a lot of dust and he deserved to be resurrected—he had a cheerful well painted face, quite a distinguished little painting, which should bring some style to our room.

Far more exciting though, amongst the paintings in storage, was the second sitter—Sir James Dewar, Fellow of Peterhouse, Fullerian Professor at the Royal Institution in London and Jacksonian Professor at Cambridge. He taught at Cambridge from 1875 and was responsible for a series of inventions, one of which involved obtaining liquid oxygen and hydrogen. The double-walled vacuum flask, represented in this canvas and named after him, was used in all chemistry laboratories around the World. This is the ubiquitous 'thermos' flask, more politely known as the 'Dewar flask'. The painting had an uncanny resemblance to my grandfather's Victorian photograph, in his apothecaries laboratory in Buzãu, and I took to it immediately.

In Romanian textbooks Dewar was revered as a world scientist and I was shocked to find him relegated, by some unseen if silent 'cultural revolution', to a corner of a College cellar. True, Dewar had not endeared himself in Cambridge, where he indulged in severe bullying of his staff.

Neither did he form a 'school' around himself, although his early experiments in low temperature physics (the liquefaction of hydrogen in 1898), preceded by a good generation work done by Rutherford's pupils on the liquefaction of helium, which he tried but did not succeed. To this continental scientist, Dewar was an icon and his abandonment to a college repository was an unmitigated piece of iconoclasm.

Sir William Quiller Orchardson, RA (1832–1910), who painted this portrait, was better known for his painting *Napoleon on Board the Bellerophon*, hanging in the Tate Gallery and other works in the 'Uffici' of Florence. Dewar's portrait was an important work, which was probably commissioned soon after the scientist was knighted in 1904. The style of the brush prefigured the Edwardian taste and the chromatics of the portrait was characteristic of Orchardson's muted colours, of predominant yellow and browns. I felt a sense of urgency in rescuing Dewar's rubicund face straight away and bringing it to light. Today it still hangs in its rightful place, amongst research students, in the GCR at Peterhouse

HENRY MOORE

My encounter with Henry Moore goes back to the artist's Retrospective Exhibition, organized by the British Council in Bucharest in the mid 1960s. It was Nikita Kruschev's dismantling of Stalin's Cold War vestiges that enabled countries in the West to show some of the best examples of their art in the East, and Henry Moore was part of this strategy. I was therefore deeply moved to see some of Moore's sculptures at Cambridge, decorating the open spaces of Churchill College.

Churchill was one of the more recent foundations in Cambridge, built somehow geographically eccentric, on a site off Madingley Road. It had the classical plan of rectangular courts, typical of the older colleges, yet the plan was more 'open' by the use of pillars, under which one could walk from one court to another. The local Cambridge yellow brick was predominant and made the three-storey buildings lighter still. The quadrangles had their indispensable manicured lawns, punctuated by contemporary sculpture. In such surroundings, the *Reclining Figure* of Henry Moore added grace to the buildings and blended in perfectly.

When visiting friends in Churchill, I often looked with delight at the inspired marriage of modern sculpture and architecture. During the hot summer evenings of East Anglia, when the undergraduates with their pints of beer spilled over from the JCR into the space of the quadrangle, I looked in bewilderment as some of them left the empty bottles and glasses on the pedestal of the *Reclining Figure*, an act of iconoclasm, I thought. Later on I came to realize that this was nothing as sophisticated as that, it was merely taking for granted the comforts of one's surroundings, treating the extraordinary with ordinariness:

"It is there" sort-of-thing attitude, "who cares?"

Would the sculptor have approved of this treatment? Maybe he would have been pleased to mix in everyday life and be accepted as an *ordinary object*, rather than be idolized, a transcendental form of recognition from the *singular* to the *omniscient*.

THE YALTA SCRAP OF PAPER

Churchill College was most famous for being the depository of Churchill's archives, which were kept in its library. This was open to researchers and scholars. As contemporary history was not my particular field, I never tried to gain access to any documents, although I was very interested in the history of the Second World War and read Churchill's account of this. Maybe my reticence in entering the College Library to look at Churchill's archives was coloured by the view, widely held in Eastern Europe, concerning Churchill's 'horse trading' with Stalin at Yalta, at the expense of the freedom of millions of people. As the 'Iron Curtain' (a Churchillian metaphor coined during his political speeches in America) divided the captive nations of Europe from their free sisters in the West, a whole generation grew up in darkness, the outcome of a deal scribbled on a scrap of paper at a dinner table at Yalta, over vodka and caviar. Churchill suggested to Stalin the partition of the spheres of influence between East and West, with percentages in each country where the Russian troops gained a foothold, or the Communists were about to take over. Poland, whose pilots fought so valiantly in the Battle of Britain and which had a legitimate government in exile which was anti-Communist and anti-Russian, was entirely abandoned to Stalin (90% Russian influence and only 10% for the West). The case of Greece was quite the reverse, where the Communists were sufficiently strong to take over the whole country, yet Churchill suggested on paper at Yalta, only 10% influence to the Russians and 90% to the West. Was this political strategy meant to ensure links with India over the eastern Mediterranean? Or rather, was it romanticism induced by Byron over classical Greece, taught in every public school? At Harrow, Churchill learned the classical Greek texts, which must have deeply influenced his judgement: Greece was the 'fountain of ancient civilization, before Rome' and could not be left under the boot of barbarian hordes!

When, soon after the war, the old man understood that Stalin had cheated him, it was too late to change the balance of power. The likeable 'Uncle Joe' soon displayed his true mettle, as a ruthless dictator from the Caucasus, no partner to the civilized Churchill. What a bitter lesson and what a price to pay! In the meantime, the fateful scrap of paper to which Churchill referred in his *Memoirs* was sitting dutifully catalogued in the archives of Churchill College, a witness to what all Eastern Europeans felt as a betrayal.

One day, the *Cambridge Evening News* splashed as a sensational story on its front page the disappearance of the dreadful scrap from Churchill's archives. Who could have stolen it? Should one rejoice, rather than shed a tear over its disappearance? What would it matter? The dreadful deed was done, with half of Europe being enslaved to the Russians at the stroke of a pen.

ROMANIAN POETRY EVENING

What better place to lament the state of Ceauşescu's Romania than at Churchill's College? With the help of Timothy Cribb, a Fellow of Churchill College, I had translated into English a series of contemporary Romanian poems, by Marin Sorescu, which I had published in *Encounter*.

Four people recited the poems: Timothy Cribb from Churchill, myself, Ben Knights from Peterhouse and Ben's girlfriend, a student of English. The poems were recited against a background of panpipe laments or *doina de jale*, Carpathian shepherds songs, which were eminently suitable for the poet's outcry, such as in my translation of the following verse:

Passport

To cross the border
Between the sunflower
And the moonflower
Between the alphabet

Of hand-written events
And printed events.
To be friend of all atoms
Which form the light
To sing with the atoms which sing
To cry
With the atoms which die
To enter into all the days of one's life
Without restriction
No matter whether they fall on one side or other
Of the word
Earth.

This passport
Is written in my bones
In my skull, femur, phalanxes and spine
All arranged in a way
To make clear
My right to be man.

(*Cambridge Review* 92, 2203, p 233, 28 May 1971)

The show took place in the auditorium of Churchill College, which was full with a young audience, attracted by the popularity of George Steiner, who agreed to make an introductory speech. George Steiner was later to have a Chair of Comparative Literature created for him at Oxford, whilst Marin Sorescu, in the post-Ceaușescu times, became Minister for the Arts in Romania and rather more conformist than in his younger days, when his poems struck an unusual mild note of dissent.

CULTURAL REVOLUTION

In the early seventies, Romania was undergoing its own 'Cultural Revolution', modelled after China where the great *Conducător* went on a state visit. Ceaușescu was so impressed with the Chinese experiment, that he declared it a worthwhile exercise to be meted out to the Romanian people.

Nobody in England heard of Romania's absurd excesses, which took another twenty years to catch the World's sympathy. However everybody was up in arms about the Chinese Cultural Revolution and the excesses of the Red Guards, who hung the resident cats of the British Embassy in Peking, not the 'fat cats', metaphorically speaking, just the feline variety. This had inflamed the indignation of the British people, a nation thriving on pet culture, whether represented by biscuits, tin foods, baskets, collars, vets, or souvenirs, quite a prosperous industry!

In a loveless society, the pet had a therapeutic part to play, especially with the children and the elderly, or even the single parent. Being a dog in England was no mean thing and I would have rather had a dog's life in Britain than a professional's existence in Eastern Europe. Well, this is not to admit that I would be an 'economic refugee', perish the thought! Here the dogs would be allowed even to lift their legs against an oeuvre of Henry Moore: this *is* real freedom! Here, at least I could empty my pint of lager seated against a bronze of Barbara Hepworth, or rub shoulders with Lurçat.

In the middle of all this public hysteria raised over the Peking cats, Joseph Needham, the unconventional Master of Gonville and Caius and the undisputed expert on the history of Chinese science and a sinologist, struck a singular note. He unashamedly loved China, when everybody else hated it. Even in a free country, such as Britain, departure from the norms was frowned upon. Worse, as a contemporary of Philby and Maclane, Needham was looked upon with greatest suspicion by the more conservative dons, as he was of the vintage of Cambridge luminaries who produced more than one spy for Soviet Russia.

Joseph Needham once told me about the reaction of a visiting Chinese official, who, on being confronted in Cambridge with reproachful remarks on the Cultural Revolution, quipped:

"But surely, you English lived your own Cultural Revolution, some time ago, under Cromwell."

Such astute comparison, not without foundation, when considering the damage inflicted by Cromwellian troops on English cathedrals, left his audience speechless.

KETTLE'S YARD

Beyond the Backs, strolling from Trinity and St John's towards Magdalene, a picture postcard view of Cambridge could be seen in Northampton Street. The cottages are set back from the street, behind a spot of triangular green and the spire of the old St Peter's church in the background. It is more akin to a corner of a rural village, rather than a place in town.

Kettle's Yard is a small cluster of small buildings of mellow brick under a pitched roof of yellow-orange, hand-made tiles. The cottages are linked together to form a single living space, yet nothing special would be thought to exist behind this elevation. The external, picturesque but modest aspect was the hallmark of the owner, Jim Ede, a former curator at the Tate Gallery, a friend of sculptors and painters and a discerning collector of modern and contemporary British art.

Jim was of a slight, almost frail frame. He could have been in his mid-sixties, or perhaps seventies, one could not tell. He was mild-mannered, softly spoken, with a permanent gentle smile, inquisitive, almost ironic eyes, floating about like a spirit of the place, radiating an immense warmth and displaying an infectious love for the objects that surrounded him.

His collection was the love of his life. Each object had a story of the encounter with the artist, an anecdote about the genesis of the particular painting or sculpture. Jim Ede befriended the artists when they were not yet recognized by the museums and the general public: he could see the potential, not as an investor, or art dealer in objets d'art, but as a great believer in the intrinsic aesthetic value of the oeuvre. Ben Nicholson could not sell his paintings for years and he would let Jim have some for the price of the canvass and frame. Prominent amongst the foreign artists was a French sculptor, Henri Gaudier-Brzeska, who lived for a while in London and died very young, during the First World War. Ede had promoted Gaudier's works and helped the posthumous recognition of the artist. *The Savage Messiah* was a biography of Gaudier, written by Jim Ede and subsequently used as a script for a British film by Ken Russell. The only sculptures owned by the Musée d'Art Moderne in Paris, now at the Beaubourg, were donated to the French Nation by Jim Ede, who believed that the name of Gaudier should be recognized and honoured in his native country. For this deed he was made an Officer of the Légion d'Honneur. It was not unusual for French artists to be first spotted by foreign collectors, before their fellow countrymen would acknowledge them. However, the majority of Jim Ede's collection was composed of British modern and contemporary artists.

In the early 1960s 'Kettle's Yard' was just a private collection, part of the 'Town' and not of the 'Gown', yet it was the perfect example of a place where the two sides could meet and make a contribution to each other. For us students, even those who did not read art, this was a place of immense inspiration and influence. This was a place where one could learn and appreciate beautiful things, not just oeuvres of established artists, but simple objects for everyday use, ceramic bowls, china plates, vases, glass, or simple forms and colours of nature, like pebbles, shells, drift wood washed up on the beach. It was natural that the place should eventually become part of the University, but the process was not simple, because, although Jim tried very hard to influence the administration into accepting it as a gift, this was not as straightforward a matter as might have been. Eventually, in 1966, the University accepted the benefaction.

By far one of the most daring features in 'Kettle's Yard's' activities was the subsidiary collection of prints and drawings which could be borrowed, for two weeks at a time, by the undergraduates, in order to hang them in their rooms in College. Instead of having a poster, or some reproduction of dubious value, one could enjoy looking for a little while at the work of an artist from 'Kettle's Yard' collection, chosen by Jim Ede. One would go to the store room under the eaves, choose one of the pictures from the double decker rack and put one's name down in a book, together with the name of the College, and two weeks later bring the picture back and replace it with another one.

This lending collection had an immense impact on the artistic perception of young people and perhaps its contribution was equally important and certainly more enduring than that made by the formal lectures at University. How many of the students who frequented the genie of the place might have become collectors of contemporary art? I should think more than one and it certainly galvanized my attention to modern British Art, beyond the oeuvre of Henry Moore and Graham Sutherland, whom I knew from my Romanian days. It was a window open to a completely new world of peace and tranquillity and timelessness, all achieved through the marriage of beautiful things put together with the greatest sensitivity in a living space, which was also a home. By going there, we learned a new way of life, a new way of love and in the process, our spirit became richer.

BRANCUSI'S GHOST

Constantin Brancusi, the son of a Romanian peasant, who went on foot to Paris, having no money to take the train to France to study art, was Romania's most revered artist. He was Rodin's pupil, but soon was to break loose of his Master's influence, which he dismissed as 'beef steak', and found his own style, which was to herald the beginning of modern sculpture. My chance encounter with Alexandre Istrati and Nathalie Dumitresco, who were Brancusi's executors, allowed me to visit his atelier in Impasse Ronsin in 1968, before Montparnasse was demolished under Pompidou and the atelier was moved to Beaubourg. Quite unexpectedly, in Cambridge, I was to meet with Brancusi again.

Under the Communist régime Brancusi's work was shrouded in a conspiracy of silence, because of his abstract, therefore 'decadent' style, contrary to the tenets of the official Socialist Realism style imported from the Soviet Union. Furthermore, Brancusi chose to ignore the régime in Romania and never came to visit his homeland after the war. Soon after the sculptor's death in Paris, Kruschev started the détente, so, out of the blue, Brancusi was posthumously 'rehabilitated'. The pendulum swung in the opposite direction—after years of this shameful conspiracy of silence about Brancusi, the artist had at long last become topical in his homeland. Romanians became news-hungry of Brancusi. There was a Brancusi cult and all art historians jumped on the bandwagon. For some, it was sheer opportunism, as they never stood up to be counted when the artist lived in exile, and never defended his work in Romania. Now Brancusomania was in full swing and anything about Brancusi was newsworthy. Compared with France, or with the United States, England was a more unlikely place of contact with Brancusi and therefore I thought that the fact was worth a mention. Whilst in Cambridge, I gathered material to publish an article in the *Revista Manuscriptum* in Bucharest. *Manuscriptum* was a quarterly journal, intended for the publication of original unpublished manuscripts, relating to historical events, or to the biography of famous people, writers, poets, artists, politicians, etc. Brancusi fell into the category of an important artist and the manuscripts in question were the correspondence he had with Jim Ede. The interest was compounded by the fact that, in one of his letters, Brancusi had some sketches of a recurrent theme in his oeuvre—that of *The Kiss*. This motif was based on one of his early sculptures and continued in *The Gate of the Kiss*, which he had erected in Romania, at Târgu Jiu. These documents were more than of passing interest, they were crucial in establishing the date of genesis of a certain concept in the mind of the artist and therefore of prime importance to the art historian.

I wrote to the Editor in Bucharest, asking if he thought the subject of sufficient interest and if so whether he could secure the prerequisite space in a future issue. The answer was positive. I asked Jim to lend me Brancusi's letters. I also took a picture of Jim in front of one of Brancusi's *Prometheus* and added a period photograph taken in the sculptor's own atelier in Montparnasse, in the late 1920s. This was lent to me by Mrs Helen Herklots, a friend from Peterborough, whose father was Professor Murgoci, a distinguished Romanian geologist and a friend of Brancusi. In the process of reading these letters I discovered, with a certain surprise, that as late as 1933, the sculptor could not write French very well. Having translated the correspondence from French into Romanian, I gave an introductory comment on 'Kettle's Yard', Jim Ede and his friendship with Brancusi.

Professor Sir Leslie Martin is remembered at Peterhouse for the design of the William Stone Building, erected in the early 1960s, as the first high-rise building in Cambridge, intended to house undergraduates. Leslie Martin is better known for the Festival Hall in London's South Bank, for which he was eventually knighted to become Sir Leslie Martin. He was also Head of the Architecture Department in Cambridge.

One day I received from Romania a rather splendid collection of photographs of Brancusi's earliest works, together with the architectural ensemble at Târgu Jiu (*The Table of Silence, The Gate of the Kiss* and *The Column of Infinity*). Amongst these pictures were also examples of Romanian artisan wood carving, which inspired Brancusi.

There was already a link between Brancusi and the Cambridge Department of Architecture, as Jim Ede had established a 'Brancusi Travel Grant' for students to be assisted on their visit to the States, where they could view Brancusi's work at Guggenheim, the Metropolitan, Philadelphia Museum of Art and elsewhere. I thought, for that reason, that my Brancusi photographs might be of interest to students of architecture at Cambridge, by displaying them in an exhibition.

I knocked at Sir Leslie's door and was received with a courteous smile, without fuss or preamble, although we had not met before and he did not know the purpose of my call. Did he know about the influence of Romanian folk art on the sculptor's work and did he know Brancusi was of Romanian stock? No he did not.

Well, I could demonstrate that by showing him a selection of my black and white photos.

He promptly agreed to let me have an exhibition in the foyer of the Department of Architecture.

The exhibition had a rather unexpected reaction, from an unusual quarter: on talking on a different subject to Maurice Cowling, a historian and Fellow of Peterhouse, he advised me against my 'enthusiasm for promoting Romania'.

"But why?" I asked incredulously. This had nothing to do with politics! Brancusi died an exile in Paris and his art was universal, only his roots were specifically Romanian and the latter had nothing to do with Communism.

THE CAMBRIDGE REVIEW

I decided to take my argument into the open, as Maurice himself so often did in the pages of the *Spectator* and other Tory press. For this purpose I was content to use the *Cambridge Review*, which would have had a greater local impact and besides would be open to an airy topic on Brancusi art criticism and its underlying foundation.

Iain Wright, editor of the *Cambridge Review* was a Research Fellow of Queen's. First, I had to sell him my idea, by explaining, in a graphic form,

Brancusi's East European connection. I went to see him in his rooms in Fisher Building, off Silver Street. The subject was topical, as the Editor was planning an Eastern Europe issue. In 1971, the memory of the Russian invasion of Czecho-Slovakia was still fresh in the public's mind and Eastern Europe was not entirely forgotten. Brancusi's *Table of Silence* in his native Romania was splashed on the front page of the *Review*, with a totally different meaning from that originally intended by the sculptor. The *Round Table* was the Soviet Union and the chairs surrounding it were the satellite states; still a Cosmogonic symbol, I should say, as originally intended by the artist.

The message, which I was trying to convey was the content of the Romanian myth in Brancusi's oeuvre, as indicated by the title of my article. My coded message to Maurice was taken from an excerpt from *Encounter*, by Claude Levi-Strauss:

"You cannot make mythology understood by anybody of a different culture, without teaching him the rules and the particular traditions of that culture."

This quote I used as the motto of my feature article, which was richly illustrated and got a prominent place in the *Review*.

My English friends in Cambridge were particularly pleased about my foray into the realm of history of art, especially since I was a geophysicist by training. It required some guts to perform such an act, in an area outside one's subject, almost as if one performed a striptease in public.

"What if one made a fool of oneself?"

This was the prevailing inhibiting factor amongst my contemporaries, who would not dare publish in a subject other than their own and even then with the utmost care. Personally, I had no such qualms, I went straight to the charge and was ready to suffer the consequences, should there be any. Here the 'consequences' were most favourable and I could see an almost conspiratorial twinkle of encouragement, in the eyes of acquaintances and friends, for my daring exhibitionism:

"Now we know what you are up to!"

CHAMPAGNE BREAKFAST

Molly Wisdom lived in a small Regency house, in Melbourne Walk, off the north side of Parker's Piece. The fact that her alleyway was called 'walk' meant that no vehicles were allowed along it, as it was too narrow anyway. Molly's house was squat, with a basement below street level, in front of which was a rockery garden. The bay window extended to the ground floor. There was a first floor above it. The roof was clad in slate. The entrance was quite unprepossessing, with a fanlight, painted in thick paint, and a doorknocker and slit for the mailbox. At the back there was a small, secluded town garden with lawn and herbaceous borders accessed from the basement and from the ground floor above. Molly's living quarters were in part of the basement, where she had a living-cum-dining room overlooking Melbourne Walk and the rockery and a small kitchen which overlooked her back garden. However, Molly's world was not limited to this backwater. Her net stretched far wider than this, way out into the political realm of Liberal constituency, in the Town Hall and beyond. She had many a contact in academia by virtue of working for the University, but also by being the ex-wife of a Professor of Philosophy and Fellow of Trinity College. She knew a great many people in Cambridge. Her basement dining room in Melbourne Walk was always open and there I had many a glass of sherry, talking about Cambridge politics. Her sherry decanter was always full and her listening capacity always at the maximum: she could absorb more and more.

Once a year, Molly would organize a champagne breakfast at her home in Melbourne Walk, intended for the Geophysics Department where she was Secretary. We all mucked in. This was more of a brunch than a champagne breakfast, as breakfast would merge into lunch. We would all turn up at about nine o'clock on a Sunday and leave in the late afternoon. The food was served in the garden and the basement, as there was enough room for the thirty or so guests, some of us with wives, husbands or girl-friends. It was a jolly party, unfolding gently in the lazy sun. Our hostess had a zest for life which put some of the young geophysicists to shame. Good old Molly, she will always be remembered fondly for her generous initiatives of this kind, of which we were all beneficiaries at least once a year until her retirement.

Molly Wisdom's retirement was just a freak of the calendar, which could not be circumvented by the University bureaucracy. Molly had to have an active retirement and she actually did more things than ever before. She carried on with a secretarial job and learned word processing quite early on, when computers were not terribly user-friendly and she was over sixty. She redoubled her political activity for the Liberal Party, of which she was a staunch supporter, and stood as a candidate in the local elections. Fundraising for various good causes in Cambridge and nationwide took a great deal of her time. She would write to the former geophysics students and ask for contributions when it came to a Cambridge cause. She did not omit me either and how could I refuse her? The Corn Exchange was her aim at the time and she tried to persuade me by saying:

"You do not have to give very much if you want your name inscribed on a list of benefactors." How could I possibly decline? I loved Cambridge and I had a debt of gratitude to Molly!

I was inscribed in the end, partly due to Molly's persistence, partly due to the fact that a building of some architectural merit in the centre of town deserved being restored. I never had the pretence of being mentioned as a benefactor, even on some doormat at Peterhouse new Library or a boathouse. Somehow, the Latin adage rang in my memory: "nomina stultorum, ubiquae locorum"(The name of silly people all over the place).

Cambridge alumni were not adverse to having their names inscribed even on library seats (for a price). The only exception to which I acceded, at Molly's insistence, and where my name appeared on a plaque, was soon obscured, as the benefactor's names lay behind piles of racks with marketing paraphernalia for cultural events at the Corn Exchange.

Gratitude is a rare commodity, for which even vanity cannot pay the right price.

NO 3, ADAM'S ROAD

The first garden party I was ever asked to was in Adam's Road, at the Adrian's. Although I had stayed at the house during my first interview in Cambridge, I never knew the garden as an extended space for entertainment. This was quite different. Both Richard and Lucy were typical Cambridge academics, from a long lineage of Oxbridge dons. Maybe the ghost of Charles Darwin was at the garden party too, or more likely his descendants. However, I was completely oblivious and did not ask questions, rather answered the unequivocal quizzing, such as how I 'escaped from Romania'.

At least Romania was on the map in Cambridge, rather than 'Ruritania' as in Newcastle. Thankfully, I had not offended anybody, as I was asked again the following year.

I could not remember if one had the ubiquitous cucumber sandwiches, but what I do distinctly remember was that here the guests at least did not jostle to get to the food, as I could see was the case at many a Cambridge garden party.

No 3 Adam's Rd had become a familiar port of call. I cycled regularly past the house on the way from Peterhouse to the Geophysics Department in Madingley Rise.

It is at No 3 that I enjoyed an unlimited hospitality. For an uprooted Romanian, who hardly knew anybody in town, this was extremely kind, especially as the Adrians had a very busy life. They were without fuss and above all natural listeners. The latter I desperately needed, as I had a lot to unravel in my early days, when I absorbed so much and understood so little. Some helping hand from a friend, a useful hint, a second opinion were all worth their weight in gold.

"Would you be free to come to dinner tomorrow? It will be just the three of us."

"Of course, I will be only too delighted."

On other occasions I would just say 'hello', hopping off my bicycle on the way to and from the Geophysics Department.

Richard's field of research in physiology dealt with the properties and activation of muscle and the mechanism that initiates movement, as well as the electrical properties of living cells.

At the time of my troubles with my first supervisor in Cambridge, Richard was a Fellow of Churchill College, where Sir Edward Bullard was also a Fellow. They would meet at Governing Bodies, High Table or other College functions and I am sure that without having been asked at all, Richard must have warned Teddy about my difficulties. He had a natural sense of civic duty, always ready to give a helpful hand, in a tactful way.

"Are you looking forward to becoming a member of the House of Lords?" I once asked Richard.

"When my father dies, I hope to take an active part in the Upper House."

Richard and Lucy often went to 'Umgeni', their father's house at Cley-next-the-Sea, a tiny group of modest cottages spilled over a sand spit, near Holt, on the north Norfolk coast.

One weekend I was asked to join them in Cley with my fiancée Roxana. Richard came to fetch us at the nearest train station.

"How is Lord Adrian?" I enquired about the frail state of his father, with whom, unbeknown to Richard, I was exchanging greetings when I sneaked into the Fellows' garden at Trinity.

"He is in hospital in Oxford, being fixed with a pacemaker."

"How come? I only saw him last week". I could not admit where, because I should not have been there.

"Yes, last week he was all right. But he got into the habit of driving over to Oxford and on the journey he must have had a heart attack, which caused him to drive his car into a lamp post, in a village close to Oxford. The physical shock of the impact must have revived his heart, as he was perfectly all right, but he's now under observation in Oxford".

"But wouldn't you need to be near him? Do not worry about us in a case of *force majeure*: we could go back to Cambridge and you could drive over to Oxford to see your father".

Richard's sense of correctness and hospitality did not allow him to alter the plans.

"Father is well looked after. Besides he will be quite excited, as a scientist, to have this pacemaker fixed to his heart."

"I am sure he will be", I said "But will you still allow him to drive in future, even though he has his pacemaker?"

"You see, it will be very difficult to tell Father to give up driving. This is a decision he must come to by himself."

Edgar Douglas Adrian was awarded the Nobel Prize for Medicine in 1932, for his research on brain wave rhythms and related neurone discharge, measured as an EEG (electroencephalogram). This was a major breakthrough in medicine, for which he was revered in Cambridge like Isaac Newton. Indeed, it may have not been entirely fortuitous that, as an undergraduate, he lived in Trinity on E1—in Newton's old rooms.

'Umgeni' was a very modest cottage and its interior typical of that of a dedicated scientist. Perhaps Molly Butler in the passage of her book on RAB gives the best description of Lord Adrian's taste for interior design. Here, Molly mentions her visit to the Master's Lodge in Trinity, where Lord and Lady Adrian were the outgoing occupants: it was described by Molly as 'high-thinking and low-living'. As for 'interior decoration' this must not have entered the concept or the vocabulary of a Nobel Prize winner or that of his wife. Lord Adrian's home was a family home, whose simplicity, in its style of life, conveyed it great dignity.

SEDGWICK CLUB DINNER

Sedgwick's name is that of the most revered British geologist. Born in Yorkshire and educated as a mathematician, he revolutionized the science of the Earth by introducing, for the first time, the term 'Cambrian' for rocks older than 500 million years, which he found in Wales and named after the old name of the province: Cambria. Together with Murchison, Sedgwick also coined the term 'Devonian', following his fieldwork in Devon. Sedgwick made geology a respected subject in Cambridge, where he was appointed Woodwardian professor, and was elected President of the Geological Society in London. He had as a field assistant in Wales a young Cambridge graduate, Charles Darwin, who subsequently sent Sedgwick rock samples and fossils from his voyage to South America. Sedgwick died in Cambridge in 1873 at the age of 88, and some 30 years later the Sedgwick Museum of Geology was founded by the University as a memorial to him. Sedgwick's name is understandably present in Cambridge in a variety of places: apart from the Sedgwick Museum, there is a Sedgwick Avenue and a Sedgwick Club. The latter has membership amongst the students and staff of the Earth Science Department. It organizes scientific talks and holds regular dinners, which for some reason, at my time at Cambridge, were taking place at Emmanuel College.

The Sedgwick Club Dinner was given in the former Dominican Friary church, converted to a College Hall after the dissolution of the Monasteries. Emmanuel was founded by a Chancellor of the Exchequer in 1584, under a royal charter from Queen Elizabeth I. The Hall was remodelled late in the 17th century and further updated by Essex in the 18th century. Its present rich decoration, together with the glittering occasion of a black tie event such as the Sedgwick Club Dinner, must have caused the many Puritan forerunners of the College to recoil in their graves. During the Commonwealth Emmanuel was reputed for its strict ascetic rules, which required the suppression of 'playing, feasting, talking', that 'idle gossip of youth', which was 'a bad habit for young minds'.

One such mind was John Harvard, later to migrate to Massachusetts Bay in 1638, where he bequeathed his library and half of his estate to the University which now bears his name. Emmanuel had a long list of Puritans who migrated to New England, which makes the College much visited by our colonial cousins. These have acknowledged the trans-Atlantic link with generous bequests, amongst which is a memorial window to Harvard in the chapel, given by Harvard men on the occasion of Emmanuel's College tercentenary in 1884.

COLLEGE HISTORY

My interest in architecture caused me to explore as much of Cambridge as possible. This soon proved to be invaluable, as I managed to enrol as a guide with a local tourist firm in Cambridge, showing visitors around.

One of the most exciting tours was that of the Electricity Board and of the Transport Trade Union, on a conference in Cambridge, during Easter. At that time trade unions were all powerful and the Heath Government was at odds with the miners and dockers. The trade unionists were aggressive and very resentful of what they perceived Cambridge to be: 'a hot bed of Conservatism and privilege'.

"What? For the privilege of walking on the Old Court lawn or for saying prayer in Latin, before dinner, in Hall?"

I immediately felt a sense of belonging and went for a counter-attack, in my best, unashamed Continental style. I knew that the dockers were exceedingly well paid and that they would not have liked to swap their wages for the meagre Fellowship of a Cambridge don: they could not. Neither could they shut me up. I was holding my ground, I was fierce and could not care less what they thought. I believe that they liked my daring them. The bosses were the ones who actually liked it best. I did not know who they were, until they came over and congratulated me for speaking out:

"We liked the way you said it. Well done!"

The money for sightseeing in Cambridge kept me ticking over, without taking too much of my time. It was fun!

One idea led to another and I soon discovered, during my tours, that some of the Colleges, such as King's, Trinity or St John's, which attracted a greater number of visitors had prepared little pamphlets with the history of the College. At Trinity, G M Trevelyan, a former Master, had produced a charming little pamphlet for visitors. He was held in high esteem by historians, who generally regarded him as a '20th century Gibbon', for his invaluable *English Social History*, which had pages of a literary beauty, reflecting the workings of a poet's mind. He had a deep love for his college, to which the King appointed him, on Churchill's recommendation and he found it useful to produce this unprepossessing small college history for visitors, which was a little gem. I felt that Peterhouse deserved to have one equally well informed:

Surely I could be one up on Trevelyan by producing an illustrated guide?!

I talked to Tim Horne, a gifted amateur photographer, who was a chemistry undergraduate in Peterhouse. We both agreed that the project had to be done. Tim Horne took a series of beautiful black and white photographs of architectural details, which I deemed to be the most important in the College. The text was prepared in French, covering mainly the history of architecture of the College. It mentioned various personalities of national repute, such as Lord Kelvin the physicist, the poet Grey, the composer Thomas Campion, Babbage the mathematician and inventor of the first computer, Sir Frank Whittle, the inventor of the jet engine, all of whom were at Peterhouse. The text was translated from French into English and the guide presented as the first bilingual edition in Cambridge. It filled a gap in the Cambridge guidebook literature and I was pleased to be told, some twenty years on, that the pamphlet was reprinted in a second edition.

GOODBYE TO ALL THAT

Before the end of my stay in Cambridge, I wanted to give a farewell party to say a big 'thank you' to all those who had helped me in my trials and tribulations, who had been supportive and steadfast, who had listened and encouraged, who participated and got involved.

I was financially drained, but I knew that I could not delay the occasion. I was not certain how much longer I was going to stay and what if my exit was too hurried or undignified?

What if I became all of a sudden too absorbed by my next move and failed to thank properly? Better hold my farewell party now!

My Tutor might have raised his eyebrows as he had just managed to obtain for me a student loan to help me complete my dissertation and here I was, wishing to throw a party:

"Typical Constantin, full of paradoxes and brinkmanship."

After four years of research in Cambridge, it was not easy to uproot myself from an environment I had grown to love: I wrote a big thank you letter to Teddy, my staunch supporter. To this he responded in his inimitably generous, if touching style:

"I was so glad to get your letter and to know that I had been helpful. Of course it must be a wrench for you to leave Cambridge and to stop being a student and to take wider responsibilities, but you know as well as I do that it is time to do so. If you find at some future time that you want to come back to academic work, then no doubt you will be able to. Come and see me any time you want to. All best wishes, yours, Teddy"

TEA AT THE MASTER'S LODGE

Long after I left Cambridge I got an unexpected invitation to have tea at the Master's Lodge at Peterhouse. This came as a surprise as, during my time as a research scholar, Professor J Graeme D Clark, Master of Peterhouse, was too busy an archaeologist, collecting honorary degrees, to allow much time for entertaining the students. He knew exactly who I was and we met 'on business' at the time when I rescued Dewar's portrait from the College cellars. He was a great amateur art historian and, as a friend of Dame Elizabeth Frink, he had a number of the artist's smaller sculptures in the Master's Lodge. The occasion of this sudden invitation was my parents' visit to England, at the request of Margaret Thatcher and Harold Wilson, who both interceded with Ceauşescu to allow them to visit their family in England. This was the result of the Helsinki agreement, as both the British Prime Minister and the Leader of the Opposition tried to impress on the Romanians that they had better honour the document they signed on human rights. My parents were a test case and their visit made news. I recall the irony of Maggie's letter of thanks to Ceauşescu, after allowing my old parents to visit Britain, when she reassured the Dictator that the Romans 'will enjoy a happy family Christmas reunion', as if the hardened Communists cared about such humanity.

So the Clarks, having heard that the Roman 'cause célèbre' had not lost its momentum, quite the contrary, thought it fit to acknowledge it by asking us to come over to have tea at the Master's Lodge. As the Romans came *en famille* with grandparents, son, daughter-in-law and three-year old granddaughter, the Master's Lodge of Peterhouse had not witnessed such animation for a considerable time. It was a little like the Babel Tower, as my parents hardly spoke any English, but were fluent in Romanian, French and German. I cannot remember if Graeme Clark tried his dog Latin on Father, or whether he tried on him some halting French. Graeme offered Father one of his preprints on the abstruse subject of neolithic archaeology. Father presented him with his book on the *Electrolysis of Sodium Chloride*. It was like a dialogue in one of Eugene Ionesco's plays in the Theatre of the Absurd.

The highlight of the visit occurred when my hyperactive daughter Gisèle, who was three years old at the time, managed to spill over a pot of coffee on the cream-coloured carpet, in the sitting room.

"Not to worry", Molly Clark tried to put us at ease, "This is only the College carpet."

True to their ancient tradition, the Romans left an enduring mark of their passage in Cambridge.

POSTFACE

When on the fateful Christmas Eve of 1989 Ceauşescu and his consort were put down by a Communist kangaroo court in Romania, I cried bitterly. Yes I cried, believe me, NOT out of any sentiment of compassion for the fallen dictator, but rather out of an uncontrollable sentiment of relief that suddenly the unthinkable had happened. I never thought that I would ever set foot in Romania again; I could not do it when my parents died in Bucharest, to lay their bodies to rest, because I refused to give up my Romanian nationality, which I considered stubbornly an unalienable right of birth. I had to pay, instead, the heavy penalty of exile.

Suddenly, old sequences from my Romanian past started flooding my vision, blurred by bitter tears, almost like the memories of the dying—I had left Romania some 21 years previously and the door shut firmly behind me, like a prison door. I should have felt elated, as indeed I was at discovering a dream come true, being a 'free man in a free country', a citizen of the world. But in the heart of my heart, whilst I lived in England I constantly hankered after Romania, whether by publishing translations of poetry, organizing an exhibition or a festival, publishing an article on Brancusi, or, indeed, researching the Carpathian earthquakes whilst at Cambridge and later on advising Shell, BP, Exxon, Mobil or Amoco about the hydrocarbon potential of Romania. My life was a series of superficially contradictory incidents, yet, on closer scrutiny, I could say they were all a series of perfectly harmonious events, like the personae of dizygotic twins.

In retrospect, many of the twists and turns, which form the canvas of this book had ironic outcomes: at Cambridge I applied for jobs with many oil companies which rejected my applications, only to end up, a decade or so later, acting as a consultant to the same companies which used my services by paying substantially more for my expertize than they would have done to me as a salaried person. In at least one other instance, my unwillingness to join a white collar trade union, which weighed against my application, was eventually vindicated as the 'closed shop' was declared illegal by the British Government. Some of the companies which declined my applications were either taken over or went bust. That is not to imply that should I have joined the staff their fortunes might have changed—not at all, but I could not resist thinking that their very narrow criteria of selection had inevitably contributed to their demise.

There is no doubt that in Romania I did not fit into a mould, whether academically or politically, and it has been lucky that the chain of events led me to carry out my research under Sir Edward Bullard, without whose influence and free style my spirit would not have flourished. The same 'Cavendish style' in research leadership was evident in Runcorn's School of Physics in Newcastle, without whose help and encouragement I could not have embarked on a doctorate.

As is the case in all 'hot houses' of talented people, on occasions I inevitably encountered resistance to my forward drive and this only steeled my resolve to succeed. Without such impediments and frustrations I may have indulged too much in idle navel-gazing, for which I had little time, as I had to defend the barricades. Several of my research articles were the outcome of such momentous crises. Likewise the painful, if tedious process of circumnavigating the bureaucratic maze, whether in Romania, France, or Britain, brought me in contact with a variety of inspiring people, some of whom became staunch friends, who helped the Roman *cause célèbre* hoist its flag on higher summits: if Rutherford's boys could do it, so the Bullard boys would do it. This was a true initiation to what 'Rab' Butler described as 'The art of the possible' and which Arnold Goodman had brought to a fine tuning, amply demonstrated in my case. Such a school of life compelled me to succeed in areas where I was least familiar and had to learn the hard way, simply because there was no turning back in a one-way street, a syndrome graphically described by Kapitza in the symbol of his 'stiff-necked crocodile', on the Cavendish Laboratory.

My quixotic crusades were not a lonely road, as I had for fellow travellers a galaxy of quality men of great calibre and integrity: this was made possible only by the convergence of two most improbable spirits— the obduracy, imagination and resourcefulness of the Romanian character, grafted on the liberalism, precision and luminosity of a Cambridge mind. Thankfully, the Cambridge academic space gave ample opportunity of contacts with arts and science people alike, on a backdrop of a singularly beautiful architecture, rich in historical, artistic and literary associations. As a teenager I dreamt of becoming an architect—at Cambridge my curiosity was constantly stimulated in this direction, in a manner which made my scientific thinking more creative.

As the Iron Curtain finally came down, contacts with Romania became easier on the long path of transition. Before 1989, Ceauşescu's all-pervasive cultural terrorism imposed its strictures even on the scientific literature: few, if any, of my scientific papers published in the West could be quoted by my fellow Romanian geophysicists. Here I had to stake a legitimate claim, during a series of invited lectures, called 'The Roman Lectures', at my old school, the University of Bucharest, in 1993. This was followed, in 1998, by the publication by the Romanian Geological Survey of my entire Cambridge dissertation. The same year the University Senate elected me a Professor Honoris Causa of the University of Bucharest. Both these belated accolades are pregnant with irony, as my early student days in Bucharest were not particularly brilliant and my Romanian director of studies regarded me 'too young to study plate tectonics'. Such retrograde thinking was perfectly at home in a dictatorship isolated from the rest of the world by Ceauşescu, a cobbler by profession, or by his successor Iliescu, a Communist party apparatchik educated in Moscow: a strong breeze of fresh air was now being felt as the 1996 elections in Romania chose as President the Chancellor of the University of Bucharest, a scholar and a professor of Mineralogy, Emil Constantinescu.

Only a few years previously, Emil Constantinescu's office was ransacked by a mob of miners, rampaging through the Capital at Iliescu's request. Iliescu even thanked the miners for 'restoring order'!

After Constantinescu's election as President of Romania, I went on several private visits to see him. On the occasion when my youngest son, Vicentiu, accompanied me, we stayed the weekend with the Presidential couple at the Government official residence at Neptun on the Black Sea coast. The grandest of the cluster of government villas, once used by the Ceauşescu's and now uninhabited, was the epitome of vulgar taste, if ever there was one. By contrast, known for their more frugal taste and unfussed simplicity, the Romanian President and Mrs Constantinescu lived in a smaller villa, once used by foreign visiting dignitaries from the 'fraternal' Communist parties, such as Dolores Ibaruri, of Spanish Civil war repute, the Italian Palmiro Togliatti, or Georges Marchais, the French Communist leader. On this occasion I had the particular privilege of being allocated Georges Marchais' former bedroom. I went to bed, having said my prayers and I slept the sleep of a new-born baby, without experiencing any nightmares. That night, my Orthodox God must have been particularly receptive, as, soon after my return to England, I read Georges Marchais' obituary in *The Times*.

President Constantinescu of Romania (Professor of Mineralogy and former Chancellor of the University of Bucharest) with my son Vicentiu Roman at Ceaușescu's villa on the Black Sea coast, August 1997. (Photograph by Constantin Roman 1997.)

During the summer of 1998, as I addressed an audience of geoscientists gathered for a Black Sea Seminar at the Geological Society in London, I could not help noticing the gloom of our profession, as the price of oil was plummeting, oil giants were merging and geologists and geophysicists were again going to find themselves without a job: I urged them to go out and get (themselves) elected politicians, to become MPs and MEPs, to take control of the affairs of Britain: if Constantinescu did it in Romania, we could do it.

I felt, in a way, I turned the 'Cavendish factor' on its head:

"There is yet hope for all of us!"

Barham House, Sussex, March 1999

Suggested Reading

Earth Sciences

Brown *et al* 1992 *Understanding the Earth - A New Synthesis* (Cambridge: Cambridge University Press)

Brush S and Gillmore C S 1995 Geophysics *Twentieth Century Physics* ed L Brown, A Pais and B Pippard (Bristol: Institute of Physics Publishing)

Fowler C M 1990 *The Solid Earth: An Introduction to Global Geophysics* (Cambridge: Cambridge University Press)

Hallam A 1973 *A Revolution in the Earth Sciences: From Continental Drift to Plate Tectonics* (Oxford: Oxford University Press)

Roman C 1970 Seismicity in Romania—evidence for the sinking lithosphere *Nature* **223** 5277, 1176–8

——1973 Buffer plates, rigid plates, sub-plates *Geophys. J. R. Astron. Soc.* **33** 369–73

——1985 Buffer plates, where continents collide *The Making of the Earth* ed R Fifield (London: New Scientist Publications)

——1998 Seismotectonics of the Carpathians and Central Asia *Romanian J. Geophys.* **18** 196

Romania

Behr E 1991 *Kiss the Hand You Cannot Bite* (London: Penguin)

Drysdale H 1996 *Looking for Gheorghe* (London: Picador)

Gallacher T 1993 *Romania after Ceauşescu: The Politics of Tolerance* (Edinburgh: Edinburgh University Press)

Leigh-Fermor P 1987 *Between the Woods and the Water* (London: Penguin)

Pacepa I 1988 *Red Horizons* (London: Heinemann)

Pakula H 1984 *The Last Romantic (Biography of Marie, Queen of Romania)* (New York: Weidenfeld & Nicolson)

Roman C 1996 A mineralogist for President *Nature* **384** 300

INDEX

PLACES, EVENTS AND SCIENCE